MICROWAVES MADE SIMPLE —
The Workbook

MICROWAVES MADE SIMPLE —
The Workbook

Salvatore J. Algeri
Manager, Components Applications Engineering
Watkins-Johnson Company, Palo Alto, California

W. Stephen Cheung
Manager, Gyro Design and Manufacturing
NASA-Gravity Probe B Program
Stanford University, Stanford, California

Lawrence A. Stark
Director of Applications Marketing
Ortel Co., Alhambra, California

ARTECH HOUSE, INC.
85 Canton Street
Norwood, MA 02062

International Standard Book Number: 0-89006-204-8
Library of Congress Catalog Card Number: 86-72681

10   9   8   7   6   5   4   3   2   1

CONTENTS

PREFACE

This workbook is intended to be a tool for learning about microwave electronics and systems. It is written as a complement to "Microwave Made Simple" (Artech House, 1985) for a broad range of readers, particularly the following:

* students who wish to supplement their theoretical knowlege with practical examples.

* beginning engineers and technicians who need to develop a practical understanding of the field, and

* managers who need to develop a practical understanding of microwave topics quickly.

The authors wrote this workbook because the best way to learn new technical skills is by solving practical problems in the field. The book consists primarily of examples and exercises. By solving the problems and comparing the results with the correct answers, the reader can test his or her understanding of the subject. Some problems, labeled "Note: more difficult", will challenge technically sophisticated readers.

The book stands by itself, although much fo the material and skills required to solve these problems are covered in "Microwave Made Simple" which is cited as the reference text in this book. The first three chapters cover fundamental microwave concepts (Chapter 1), transmission line calculations (Chapter 2), and the use of the Smith chart (Chapter 3). Chapter 4 extends the use of Smith charts to the design of solid state amplifiers and oscillators.

Chapter 5 will be of special interest to new workers in the field of hybrid microwave integrated circuits (MICs). Understanding how the layout and packaging of these miniature circuits relate to the original schematic is often a major obstacle to developing confidence in the field. This chapter leads the reader through a series of increasingly complex MICs, explaining how the circuit layout is related to the original schematic.

Chapter 6 through 9 are devoted to microwave systems. The material covers noise principles (Chapter 6), antenna and propagation calculations (Chapter 7), the basics of radar systems (Chapter 8), and satellite systems (Chapter 9). Most of the material is covered in "Microwave Made Simple" but some new material has been included.

All through the book, the reader is encouraged to work out exercises to test his or her understanding of the principles which have been presented. We feel that the reader who can solve these exercises will have a good grasp of the most important concepts in microwave electronics and systems.

The authors would like to acknowledge Frederic Levien who gave us permission to use his radar problems from "Microwave Made Simple", (Chapter 14). We also acknowledge the generosity of Watkins Johnson Co., Palo Alto, CA for the information about the circuits presented in Chapter 5. The authors are grateful to Annette Cheung and Shan-Shan Ho for their help in preparing the manuscript.

S.J.A.
W.S.C.
L.A.S.

1

MICROWAVE FUNDAMENTALS

The following equations will be used for the exercises given in this chapter.

I. For an electromagnetic wave traveling in free space, the product of the wavelength L_o and the frequency f is equal to the speed of light in free space, 3×10^{10} cm/sec.

$$f(GHz) \times L_o(cm) = 30 \text{ cm/ns} \tag{1.1}$$

II. The conversion between a power ratio R and its value in decibel (dB) is

$$R \text{ in dB} = 10 \times \log (R \text{ in number}) \tag{1.2A}$$
$$R \text{ in number} = \text{inv log } (R \text{ in dB}/10). \tag{1.2B}$$

See Chapter 2 of reference text.

Useful dB values are given here.

Number	1	2	3	4	5	6	7	8	9	10	10^n
decibel	0	3	5	6	7	8	8.5	9	9.5	10	$n \times 10$

III. The conversions between a power value and its values in dBm and dBW are

$$dBm = 10 \times \log (\text{milliwatt value}) \tag{1.3A}$$
$$dBW = 10 \times \log (\text{watt value}) \tag{1.3B}$$
$$dBm = dBW + 30dB \tag{1.3C}$$

See Chapter 2 of reference text.

IV. A power amplifier amplifies an input signal and gives a larger signal at the output. The relationship linking gain and input and output powers is

$$\text{Gain (\#)} = P_{out}(mW) / P_{in}(mW) \tag{1.4A}$$
$$\text{Gain (dB)} = P_{out}(dBm) - P_{in}(dBm) \tag{1.4B}$$

See Chapter 5 of reference text.

V. An attenuator reduces an input signal and gives a smaller output signal. The relationship linking attenuation, input and output power is

$$A (\#) = P_{in}(mW) / P_{out}(mW) \tag{1.5A}$$
$$A (dB) = P_{in}(dBm) - P_{out}(dBm) \tag{1.5B}$$

See Chapter 5 of reference text.

VI. Unless perfectly matched, only a portion of an input signal to a device is transmitted and the remaining amount is reflected back. The quantities for reflection are return loss (RL), reflection coefficient (Δ), and voltage standing wave ratio (VSWR).

$$RL\ (\#) = P_{in}\ (mW)\ /\ P_{refl}\ (mW) \qquad (1.6A)$$
$$RL\ (dB) = P_{in}\ (dBm) - P_{refl}\ (dBm) \qquad (1.6B)$$
$$P_{trans}\ (mW) = P_{in}\ (mW) - P_{refl}\ (mW) \qquad (1.6C)$$

The conversions among RL, Δ, and VSWR are given as follows. All of the quantities in the following formulas are in number, not dB. If a given quantity is in dB (usually the return loss), convert into number before applying the formula.

For return loss,

$$RL = 1/\Delta^2 = [(VSWR+1)/(VSWR-1)]^2 \qquad (1.7A)$$

For reflection coefficient,

$$\Delta = 1/\sqrt{RL} = [(VSWR-1)/(VSWR+1)] \qquad (1.7B)$$

For voltage standing wave ratio,

$$VSWR = (\sqrt{RL}+1)/(\sqrt{RL}-1) = (1 + \Delta)/(1 - \Delta) \qquad (1.7C)$$

Exercises

(1.1) Do the following frequency and free space wavelength conversions using Equation (1.1)

frequency f(Ghz)	free space wavelength Lo(cm)
	3.0
	4.167
	0.667
2.5	
0.5	
2.0	

Sample calculations

f = 7.2 Ghz, L_o(cm) = 30/7.2 = 4.167 cm

(1.2) Do the following number and decibel conversions using Equation (1.2). Round off values are acceptable.

Number	dB
	16
	35
	88.5
	-17
	-63
	-45
16	
500	
7×10^6	
1.6×10^{-5}	
4.0×10^{-10}	
0.007	

Sample calculation

   a. Number = 3000, dB = 10 x log(3000) = 10 x 3.5 = 35dB

   b. dB = -48dB,   Number = inv log (-48/10) = inv log (-4.8)= 1.6×10^{-5}

(1.3) Convert the following powers into dBm and dBW using Equations (1.3A) and (1.3B).

power	dBW	dBm
5mW		
35W		
6kW		
20μW		
0.5pW		
7nW		
3MW		

Sample calculation

$$20\mu W = 2 \times 10^{-5}W = -47dBW = 2 \times 10^{-2}mW = -17dBm$$

(1.4) Convert the following dBm values into actual powers in milliwatt and watt.

dBm	mW	W
19		
37		
4		
52		
-26		
-78		

Sample calculation

$$52dBm = inv \; log \; (52/10) \; mW = inv \; log \; (5.2) \; mW = 1.6 \times 10^5 \; mW$$
$$= 1.6 \times 10^5/1000 \; W = 160 \; W$$

Exercises (1.5) to (1.8) use Figure (1.1) and Equations (1.4A) and (1.4B).

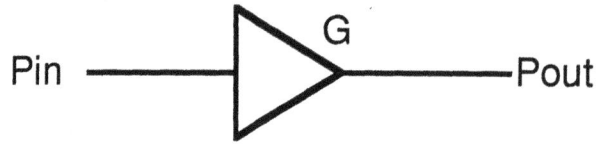

Fig. 1.1

(1.5) Given P_{in}= 20 mW (13dBm) and G=3000 (35dB), find P_{out}.

    Numerically, P_{out}= P_{in} x G = 20 mWx 3000 =60000mW =60W
    In decibel, P_{out}(dBm) = P_{in}(dBm) + G(dB)= 13dBm + 35dB= 48dBm.

(1.6) Given P_{out}=3mW (5dBm) and G=500 (27dB), find P_{in}.

(1.7) Given P_{in}= 3mW (5dBm) and P_{out}=600mW (28dBm), find G.

(1.8) Given P_{in}= 20pW(-77dBm) and P_{out}= 3µW (-25dBm), find G.

    Exercises (1.9) to (1.11) use Figure (1.2) and Equations (1.5A) and (1.5B).

(1.9) Given P_{in}= 2mW(3dBm) and A= 200 (23dB), find P_{out}.

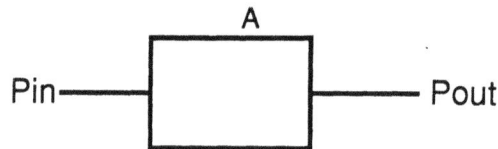

Fig. 1.2

    Numerically, P_{out}= P_{in}/A = 2mW/200 = 0.01mW (10µW).
    In decibel, P_{out}(dBm)= P_{in}(dBm) - A(dB) = 3dBm - 23dB = -20dBm.

(1.10) Given P_{out} = 4nW(-54dBm) and A=5000(37dB), find P_{in}.

(1.11) Given P_{in}= 30µW (-15dBm) and P_{out}= 50nW(-43dBm), find A.

    Exercises (12) to (14) are with reference to Figure (1.3).

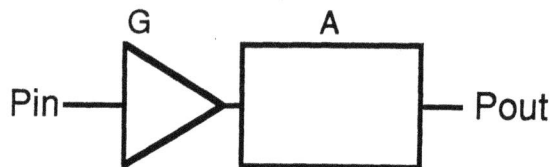

Fig. 1.3

(1.12) Given P_{in}= 3dBm, G=20dB, and A=5dB, find the output power.

P_{out}(dBm) = P_{in}(dBm) + G(dB) - A(dB)= 3dBm + 20dB -5dB =18dBm.

(1.13) Given P_{in}= -25dBm, G=37dB, and P_{out}= 2dBm, find A.

(1.14) Given P_{out}= 4dBm, G=10dB, and A= 30dB, find P_{in}.

(1.15) Find the output power of the circuit in Figure (1.4) given that P_{in}= -5dBm, G_1= 25dB, G_2= 23dB, A_1=30dB, and A_2=20dB.

The output power P_{out} is given by

P_{out}(dBm) = P_{in}(dBm) + G_1(dB) - A_1(dB) + G_2(dB) - A_2(dB)
= -5dBm + 25dB - 30dB + 23dB - 10dB = 3dBm

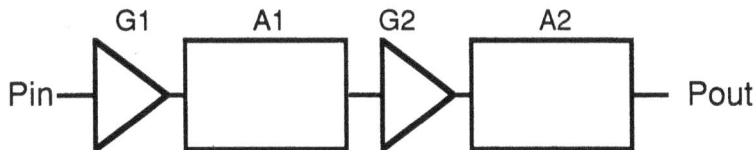

Fig. 1.4

(1.16) Three amplifiers are connected by two cables as shown in Figure (1.5A). The gains of the amplifiers are G_1=20dB, G_2=15dB, and G_3=10 dB while the lengths of the cables are l_1=10ft and l_2=20ft. The cable attenuation is 0.3dB/ft. What is the output power if the input power is -30dBm?

The cable attenuations can be regarded as real attenuators, A_1=3dB and A_2=6dB, in series with the amplifiers, see Figure (1.5B).

The output power is

P_{out}(dBm) = P_{in}(dBm) + G_1(dB) - A_1(dB) + G_2(dB) - A_2(dB) + G_3(dB)
= -30dBm + 20dB - 3dB + 15dB - 6dB + 10dB = 6dBm.

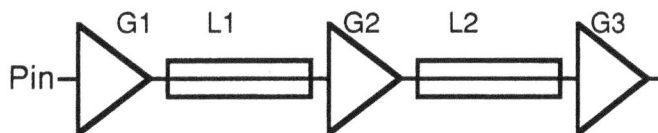

Fig. 1.5A

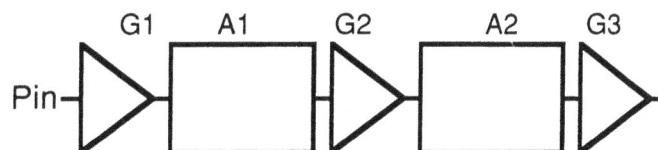

Fig. 1.5B

(1.17) Four amplifiers are connected in series with their gains given to be G_1= 30dB, G_2=23dB, G_3=15dB, and G_4=12dB. The cable lengths are l_1=10ft, l_2= 15ft, and l_3=20ft. Given that the input power is -30dBm and the output power is 5dBm, what is the cable attenuation per foot?

Let the cable attenuation be y dB/ft (y is an unknown), the attenuations of the three cables are equal to 10y dB, 15y dB, and 20y dB respectively. The unknown y can be found by solving the following equation.

$$P_{out}(dBm) = P_{in}(dBm) + G_1(dB) - A_1(dB) + G_2(dB) - A_2(dB) + G_3(dB) - A_3(dB) + G_4(dB).$$

$$5 = -30 + 30 - 10y + 23 - 15y + 15 - 20y + 12$$

i.e.

$$5 = -30 + 80 - 45y$$
$$45y = 80 - 35 = 45$$
$$y = 1.0$$

The cable has an attenuation of 1.0 dB/ft.

(1.18) Four amplifiers are connected by three cables. The gains of the amplifiers are G_1=12dB, G_2=23dB, G_3=16dB, and G_4=5 dB while the lengths of the cables are l_1=25ft, l_2=10ft, and l_3=35ft. The cable attenuation is 0.2dB/ft. What is the output power if the input power is 0dBm?

Exercises (1.19) to (1.21) are referenced to Figure (1.6) and Equations (1.6A) through (1.6C).

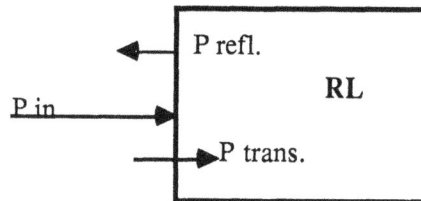

Fig. 1.6

(1.19) A generator is connected to a mismatched device. The input power to the device and the reflected power are measured to be 10mW(10dBm) and 2mW(3dBm) respectively. What is the return loss of the device? How much power is transmitted and what is it as percentage of input power? What is the mismatch loss?

Numerically, RL(#) = P_{in}/P_{refl} = 10mW /2mW = 5.
In decibel, RL(dB) = P_{in}(dBm) - P_{refl}(dBm) = 10dBm - 3dBm = 7dB.
Power transmitted P_{tran} (mW) = P_{in} (mW) - P_{refl}(mW) = 10mW - 2mW = 8mW = 9dBm.
Percentage of transmitted power = 100% x P_{tran} /P_{in} = 100% x 8mW/10mW = 80%.
Mismatch loss = P_{in}(dBm) - P_{tran}(dBm) = 10dBm - 9dBm = 1dB.

(1.20) The return loss of a device is 25dB. Find the reflected power for an input power of 6dBm.

(1.21) The power of the reflected signal from a device, return loss 35dB, is measured to be -48dBm. What is the input power?

(1.22) An isolator is the microwave analog of a diode, its attenuation in the forward direction is small, while the attenuation in the reverse direction is very large. An isolator connected to a signal generator can protect the generator against the undesired reflected signal from the remainder of the circuit. The forward loss and the isolation (reverse loss) of an isolator are 0.5dB and 30dB respectively. With reference to the circuit given in Figure (1.7), the output end of the isolator is shorted so that all the forward signal is reflected back. Find the power of the reflected signal to the generator if the generator output is 10dBm.

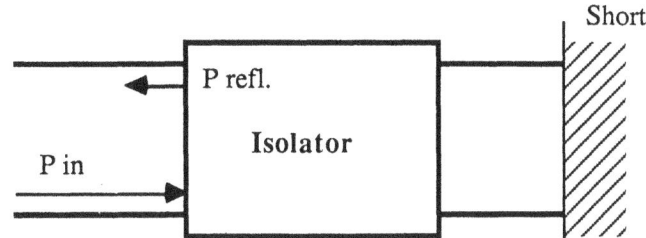

Fig. 1.7

Let P_{in} be the power of the signal in dBm of the generator driving the isolator. In the forward direction, the signal at the output of the isolator is P_{in} - 0.5dB. This signal is 100% reflected by the short. In the reverse direction, this reflected signal becomes (P_{in} - 0.5dB) - 30dB = P_{in} - 30.5dB. Therefore, if P_{in} is 10dBm, the reflected signal to the generator is 10dBm - 30.5dB = -20.5dBm.

(1.23) The short in the previous question is now replaced by a device whose return loss has been measured separately to be 20dB as shown in Figure (1.8). What is the power of the reflected signal to the generator? Also, what is the effective return loss seen by the generator?

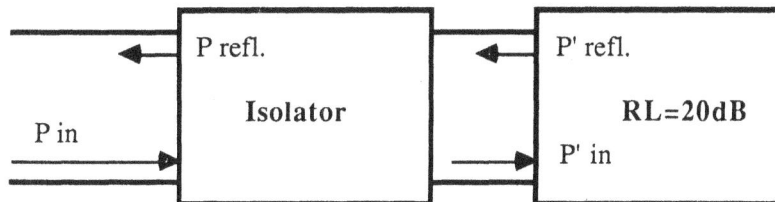

Fig. 1.8

As shown in the previous question, the signal at the output of the isolator is P_{in} - 0.5dB. This is also the input signal to the device. The reflected signal from this device is P'_{refl} = (P_{in} - 0.5dB) - 20dB = P_{in} - 20.5dB. In the reverse direction, this reflected signal becomes P_{refl} = (P_{in} - 20.5dB) - 30dB = P_{in} - 50.5dB. Therefore, the reflected signal seen by the generator with a power of 10dBm is 10dBm - 50.5dB = -40.5dBm.

The return loss seen by the generator is P_{in}(dBm) - P_{refl}(dBm), i.e., 10dBm - (-40.5dBm) = 50.5dB. Note that this return loss is simply the decibel summation of the forward loss, the isolation, and the return loss of the device, i.e., the return loss seen by the generator is 0.5dB + 30dB + 20dB = 50.5dB.

(1.24) The isolator in the previous question is replaced by an attenuator of 15dB attenuation as shown in Figure (1.9). Describe the difference in performance.

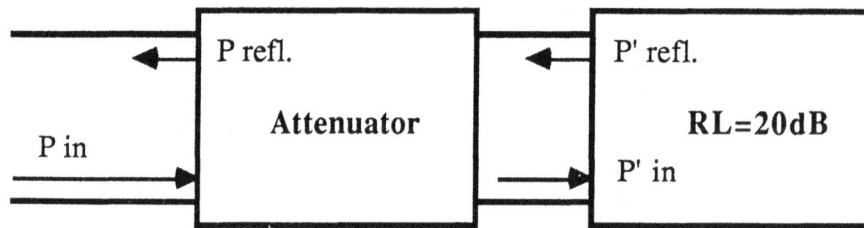

Fig. 1.9

An attenuator attenuates a signal in both direction (similar to a resistor). If the generator power is 10dBm, the power coming out of the attenuator is 10dBm-15dB = -5dBm.

This -5dBm signal then encounters the device with 20dB return loss. The reflected signal is -5dBm - 20dB = -25dBm. The reflected signal now suffers a 15dB loss, i.e., its power is -25dBm - 15dB = -40dBm on its way to the generator. Therefore, the generator sees a -40dBm reflected signal. The return loss seen by the generator is 10dBm - (-40dBm) = 50dB.

Although the return loss is almost the same as the isolator case, the difference in performance is due to sacrafice of power to the attenuator in both directions and to the device. The power available for the devices following the attenuator is much less than that following the isolator. Therefore, an attenuator can act as an isolator in protecting the generator but cannot supply the same amount of power to the device. However, an attenuator is frequently used in broad band matching due to the frequency limitation of an isolator.

(1.25) An isolator is placed between a generator and a device. For each of the following combinations of isolator forward loss, isolation, and device return loss, find the effective return loss seen by the generator. All the quantities are in dB.

Isolator forward loss	isolation	device RL	effective RL
0.2	25	30	
0.6	35	10	
1.5	30	10	

(1.26) Complete the following conversion table for return losses, reflection coefficient, and VSWR using Equation (1.7).

RL(dB)	RL(#)	Δ	VSWR
5			
	300		
			1.5
		0.8	

Sample Calculation

Given RL = 5dB=3.

$\Delta = 1/\sqrt{3} = 0.58$,   VSWR= $(1 + 0.58) / (1 - 0.58) = 3.74$

(1.27) The coupling coefficient of a directional coupler are 20dB. Find the power at the ouput and the secondary arm for an input signal of 5dBm .

The power at the secondary arm is

$$P_2(out) = P_{in}(dBm) - \text{Coupling Coefficient(dB)}$$
$$= 5dBm - 20dB = -15dBm.$$

The power at the main output is the numerical difference between the input power and the secondary arm. The input power is 5dBm, or 3mW, and the power at the secondary arm is -15dBm, or 0.03mW. Therefore, the power at the main output is 3mW - 0.03mW = 2.97mW.

2

TRANSMISSION LINE THEORY

2.1 DEFINITIONS

δ = skin depth
r = resistirity
μ = permiability
L_o = free space wavelength
L_g = guide wavelength
k = didectric constant
L_c = cutoff wavelength
f_c = cutoff frequency
Z_o = characteristic impedance
L = inductance per unit length of cable
C = capactance

2.2 SKIN DEPTH

Skin depth is the effective penetration distance when an electromagnetic wave of frequency f encounters a metal. See Section 3.6 of the reference text. The equation for skin depth is

$$\partial(cm) = \sqrt{r}/\pi f\mu \tag{2.1}$$

where r is the resistivity of the metal in Ω-cm, f is the frequency in Hz, and μ is the permeability (i.e., the magnetic property) of the metal in Henry /cm.

For non-magnetic metals the permeability is $\mu = \mu_o = 4\pi \times 10^{-9}$ H/cm. The resitivities of common metals in Ω-cm are r(copper)=1.7×10^{-6} , r(aluminum) = 2.8×10^{-6}, r(gold) = 2.4×10^{-6}, and r(silver) = 1.6×10^{-6}.

The resistivity of copper is $1.7 \times 10^{-6} \Omega$-cm and its permeability is $\mu = 4\pi \times 10^{-9}$ H/cm. Let the frequency be f GHz, i.e., $f \times 10^9$ Hz, the equation for copper's skin depth is therefore

$$\partial(cm) = \sqrt{1.7 \times 10^{-6}/\pi f \times 10^9 \times 4\pi \times 10^{-9}} = 2.08 \times 10^{-4}\sqrt{1/f} \text{ cm}$$

$$\partial(copper) \approx 2/\sqrt{f} \, \mu m \tag{2.2}$$

Exercises

(2.1) Calculate the skin depth at 5 GHz for copper, aluminum, gold, and silver.

For copper, use Equation (2.1)

$$\partial(cm) = \sqrt{1.7 \times 10^{-6}/\pi \times 5 \times 10^9 \times 4\pi \times 10^{-9}} = 9.28 \times 10^{-5} \text{ cm} = 0.928 \mu m$$

12

Instead of calculating the skin depth for each metal, we notice from the equation of the skin depth that, for a given frequency but different resistivities, $\partial_1/\partial_2 = \sqrt{r_1/r_2}$ where the subscript 1 is the quantities of metal 1 and subscript 2 for the quantities of metal 2. We have:

$$\partial(aluminum, cm) = \partial(copper, cm) \times \sqrt{r(aluminum)/r(copper)}$$
$$= 0.928 \ \mu m \times \sqrt{2.8 \times 10^{-6}/1.7 \times 10^{-6}} = 1.19 \ \mu m.$$
$$\partial(gold, cm) = 0.928 \mu m \times \sqrt{2.4 \times 10^{-6}/1.7 \times 10^{-6}} = 1.10 \mu m.$$
$$\partial(silver, cm) = 0.928 \mu m \times \sqrt{1.6 \times 10^{-6}/1.7 \times 10^{-6}} = 0.900 \mu m.$$

(2.2) Use Equation (2.2) to calculate the skin depths of copper for the following frequencies: 1.5GHz, 2.5GHz, 5.7GHz, 10.3GHz, 15.2GHz, and 25.0GHz.

f(GHz)	Skin depth in μm
1.5	
2.5	
5.7	
10.3	
15.2	
25.0	

Sample calculation

For f= 25 GHz, $\partial \approx 2/\sqrt{25} = 0.4 \ \mu m$.

2.2   GUIDE WAVELENGTH

See Section 4.11 of the reference text.

The guide wavelength, L_g, for a coaxial cable is

$$L_g = L_o/\sqrt{k} \tag{2.3}$$

where L_o is the free space wavelength. The free space wavelength is given by $L_o(cm) = 30/f(GHz)$ according to Equation (1.1).

The guide wavelength, L_g, for a 50 Ω microstrip line is

$$L_g = L_o/\sqrt{k_{eff}} \tag{2.4}$$

where k_{eff} is the effective dielectric constant and is given by $k_{eff} = (k_1 + k_2)/2$.

The guide wavelength in an air filled rectangular waveguide is

$$L_g = L_o / \sqrt{1 - (Lo/2a)^2} \tag{2.5}$$

where L_o is the free space wavelength and a is the width of the waveguide.

(3.3) Calculate the guide wavelength in a coaxial cable for frequencies 1.0, 2.5, 4.8 , and 6 GHz. The dielectric constant, k, of the cable material is 2.2.

f(GHz)	L_o(cm)	L_g(cm)
1.0		
2.5		
4.8		
6.0		

Sample calculation

For f=1.0GHz, use Equation (1.1) to find the free space wavelength L_o = 30 / 1 = 30 cm

Use Equation (2.3) to find the guide wavelength L_g = 30 / $\sqrt{2.2}$ = 20.23 cm

(2.4) Calculate the guide wavelength for a microstrip whose substrate is alumina (k=9.8) and surrounded by air (k=1.0) for frequencies 1.0, 3.5, 5.7, and 6.2 GHz.

The effective dielectric constant for the given microstrip is (9.8+1.0)/2 = 5.3.

f(GHz)	L_o(cm)	L_g(cm)
1.0	30	
3.5	8.57	
5.7	5.26	
6.2	4.84	

(5.5) The internal width, a, of a WR90 rectangular waveguide is 2.286cm (0.9inch). Find the guide wavelengths of the signals entering the waveguide of frequencies: 10.0GHz, 9.0GHz, 8.5GHz, 8.0GHz, and 5.0GHz.

f(GHz)	L_o (cm)	L_g(cm)
10.0		
9.0		
8.5		
8.0		
5.0		

Sample calculation

For f= 9.0 GHz, L_o = 30 /9 = 3.333cm.

$$L_g = \frac{3.333 \text{ cm}}{\sqrt{1 - (3.333/2\times2.286)^2}} = \frac{3.333 \text{ cm}}{\sqrt{1 - 0.729^2}} = \frac{3.333 \text{ cm}}{0.685} = 4.869 \text{ cm}$$

Note that no real, i.e., physically meaningful, answer exists for f=5.0GHz. Explanation of this effect is found in Section 4.3 in the reference text. It is also mathematically evident that L_o at 5.0GHz is larger than 2a, so $\sqrt{1-(L_o/2a)^2}$ is the square root of a negative number and does not give a real answer.

2.4   CUTOFF FREQUENCIES

See Section 4.3 of the reference text for the rectangular waveguide in the following discussion.

If the frequency f of a signal is below the cutoff frequency f_c, the signal will not go into the waveguide. For a rectangular waveguide of internal width a, the cutoff frequency is calculated from the cutoff wavelength L_c which is

$$1/2 \, L_c = a \qquad (2.6)$$

where a is the inner width fo the waveguide.

The cutoff frequency for a coaxial cable is one beyond which an entering signal will generate complicated mode patterns of propagation and loss characteristics. It is usually a good idea to have the signal frequency stay below the cutoff frequency when selecting a cable. The principal mode of propagation for coaxial cable (and all transmission lines that work at low frequencies) is the transverse electromagnetic (TEM) mode. For a coaxial cable, the cutoff frequency is given as

$$f_c = \frac{23.6}{\pi (D + d)\sqrt{k}} \text{ GHz} \qquad (2.7)$$

where k is the dielectric constant of the cable material and D and d are the outer and inner diameter in inches.

(2.6) The inner width of a WR90 waveguide is 0.9inch. What is the cutoff frequency?

Since a = 0.9inch = 2.286cm.
L_c = 2x2.286 =4.572 cm.
The cutoff frequency is f_c (GHz) = 30 /4.572 = 6.56 GHz.

(2.7) Using the RG9 cable in the next Exercise (8) to find the cutoff frequency. We know that k=2.1, D=0.28 inch, and d=0.086 inch.

Calculation

Use Equation (2.7),

$$f_c = \frac{23.6}{\pi (0.28 + 0.086) \sqrt{2.1}} \text{ GHz} = 14.16 \text{ GHz}.$$

2.5 CHARACTERISTIC IMPEDANCE

See Sections 3.5 and 4.5 of the reference text.

The characteristic impedance of a coaxial cable is

$$Z_o = \sqrt{L/C} \qquad (2.9)$$

where L and C are the inductance and capacitance per unit length of the cable in Henry/foot and Farad/foot respectively. These quantities are also related to the dimensions of the parameters making the cable, e.g., the diameters of the inner and the outer conductors of the cable. The above equation can be rewritten as

$$Z_0 = [138 \ \log (D/d)] / \sqrt{k} \tag{2.10}$$

where D is the diameter of the outer conductor, d is the diameter of the inner conductor, and k is the dielectric constant of the material between the inner and the outer conductor.

The capacitance between two conductors is related to the conductors' geometrical distribution of electric field. For a coaxial cable, the electric field is distributed radially due to azimuthal symmetry. The capacitance in picofarad per foot is given to be

$$C(pF/ft) = 7.35k / \log(D/d). \tag{2.11}$$

The inductance is related to the geometrical distribution of magnetic field. For a coaxial cable, the magnetic field is distributed as circles about the conductors. The inductance in nano-henry per foot is given without proof to be

$$L(nH/ft) = 140 \ \log(D/d) \tag{2.12}$$

Note that in both the capacitance and the inductance, the units of D and d are not important as long as they are consistent.

(2.8) Use Equations (2.11) and (2.12) to find the capacitance and inductance for the RG9 coaxial cable.

For the RG9 cable, k=2.10, D=0.280 inch, d=0.086 inch, we have

$$C(pF/ft) = 7.35 \times 2.1 / \log(0.280/0.086) = 30.1 pF/ft$$

and

$$L(nH/ft) = 140 \times \log(0.280/0.086) = 71.8 \ nH/ft$$

(2.9) Find the characteristic impedance of a RG9 cable based on the technical data given below. Cable capacitance =30pF/ft, cable inductance =71.8nH/ft, dielectric constant of the material= 2.10, diameter of inner conductor=0.086inch, and diameter of outer conductor=0.280 inch.

Using Equation (2.9), we have $Z_0 = \sqrt{71.8 \times 10^{-9}H/30 \times 10^{-12}F} \approx 48.9$ ohm

Using Equation (2.10), we have $Z_0 = [138 \ \log(0.28/0.086)]/\sqrt{2.1} \approx 48.8$ ohm

(2.10) Use the following data for a coaxial cable to calculate the capacitance and inductance per unit length and the characteristic impedance of the cable. k=2.10, D=0.945 inch, d=0.114 inch.

2.6   VELOCITY OF PROPAGATION

See Section 4.7 of the reference text.

The velocity of propagation is

$$v = 1 / \sqrt{LC}. \tag{2.13}$$

The unit of v is meter per second if L is in Henry per meter and C is in Farad per meter.

The velocity can be alternatively calculated from the following equation

$$v = c / \sqrt{k} \tag{2.14}$$

where c is the speed of light in free space (3×10^{10}cm /s) and k is the dielectric constant of the material.

(2.11) Find the velocity of propagation for the RG9 cable.

We know that C= 30pF/ft = 98.4 pF/m and L= 71.8nH/ft = 235.5nH/m.

Therefore, the velocity of propagation, using Equation (2.13), is

$$v = 1/\sqrt{235.5 \times 10^{-9} \times 98.4 \times 10^{-12}} \text{ meter/sec} = 2.07 \times 10^8 \text{ meter/sec.}$$

Using Equation (2.14), the velocity of the RG9 cable is

$$v = 3 \times 10^{10}/\sqrt{2.1} \text{ cm/s} \approx 2.07 \times 10^{10} \text{ cm/s} = 2 \times 10^8 \text{ m/s.}$$

(2.12) Find the velocity of propagation for Exercise (2.10).

3

BASIC SMITH CHART

3.1 INTRODUCTION

The readers are referred to Chapters 7 and 8 of the reference text for detailed discussion of the Smith chart.

Impedance is given as $Z = R + jX$ ohm where R is resistance and X is reactance. For an inductor, $X = X_L = +2\pi fL$ ohm; for a capacitor, $X = X_C = -1/2\pi fC$ ohm.

Note that j is an imaginary number and is equal to $\sqrt{-1}$.

Admittance is given as $Y = G + jB$ mho where G is conductance and B is susceptance. For an inductor, $B = B_L = -1/2\pi fL$; for a capacitor, $B = B_C = +2\pi fC$; for a resistor, $G = 1/R$.

Normalization is to compare a given impedance Z against a reference impedance, usually the characteristic impedance of a transmission line Z_o. In other words, normalized $Z = z_n = Z/Z_o$. Similarly, a given admittance Y can be normalized against the line admittance Y_o ($= 1/Z_o$) and is written as normalized $Y = y_n = Y/Y_o$.

A Smith chart is designed for graphic description of the standing wave formed due to a mismatch between the load impedance and the line impedance. The normalized impedance of a load can be plotted on the Smith chart and other information such as VSWR(voltage standing wave ratio), return loss, and reflection coefficient of the given circuit can be obtained. Effects on the impedance as the line length changes can also be found. Necessary steps to match the circuit can also be obtained from movements on the Smith chart.

Exercises

(3.1) Calculate the impedance values of the combinations of components at the given frequencies, both shown in Figure (3.1). Then normalized each impedance against Z_o=50 ohm.

Calculation

a. $X_L = 2\pi \times 10 \times 10^9 \times 5 \times 10^{-9}$ ohm = 314 ohm
$Z = 150 + j314$ ohm
$z_n = 150 + j314$ ohm /50 ohm $= 3 + j6.28$

b. $X_C = \dfrac{1}{2 \times \pi \times 6 \times 10^9 \times 20 \times 10^{-12}}$ ohm = 1.32 ohm
$Z = 60 - j1.32$ ohm

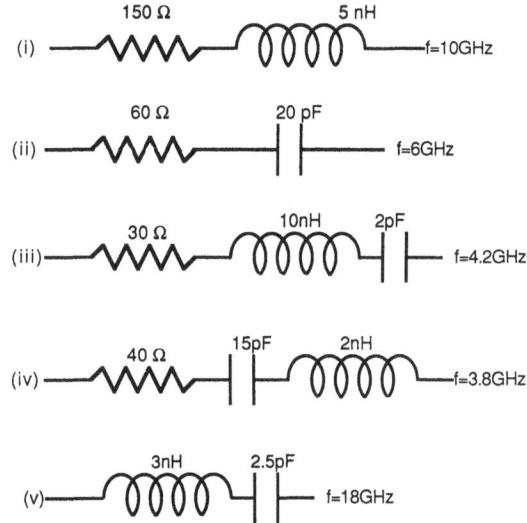

Fig. 3.1

$z_n = 60 - j1.32$ ohm $/50$ ohm $= 1.2 - j0.026$

c. $X_L = 2\times\pi\times4.2\times10^9\times10\times10^{-9}$ ohm $=263.9$ ohm

$X_C = \dfrac{1}{2\times\pi\times4.2\times10^9\times2\times10^{-12}}$ ohm $= 189.5$ ohm

$Z = 30 + j(263.9 - 189.5)$ ohm $= 30 + j74.4$

$z_n = 30 + j\,74.4$ ohm $/50$ ohm $= 0.6 + j1.49$

d. $X_C = \dfrac{1}{2\times\pi\times3.8\times10^9\times15\times10^{-12}}$ ohm $= 2.79$ ohm

$X_L = 2\times\pi\times3.8\times10^9\times2\times10^{-9}$ ohm $= 47.75$ ohm

$Z = 40 + j(47.75 - 2.79)$ ohm $= 40 + j44.96$ ohm

$z_n = 40 + j44.96$ ohm $/50$ ohm $= 0.8 + j0.899$

e. $X_L = 2\times\pi\times18\times10^9\times3\times10^{-9} = 339.3$ ohm

$X_C = \dfrac{1}{2\times\pi\times18\times10^9\times2.5\times10^{-12}}$ ohm $= 3.54$ ohm

$Z = 0 + j(339.9-3.54) = 0 + j336.36$ ohm

$z_n = 0 + j336.36$ ohm $/50$ ohm $= 0 + j6.73$

(3.2) Trace in detail the location of the normalized impedance points 0.7+j1.0 (point A), 2-j0.2 (point B), 1.0+j2.2 (point C), and 0.2-j0.6 (point D) on a Smith chart.

Calculation

The four points are plotted on a Smith chart as shown in Figure (3.2). We will describe point A in detail. Locate the 0.7 resistance circle and the 1.0 reactance curve. These two curves intersect at a common point which is point A.

(3.3) Normalize the following impedance points against $Z_o = 50$ ohm and plot them on the Smith Chart (Figure (3.3)). Indicate the nature of the impedance, i.e., resistive (R), inductive (L), capacitive (C).

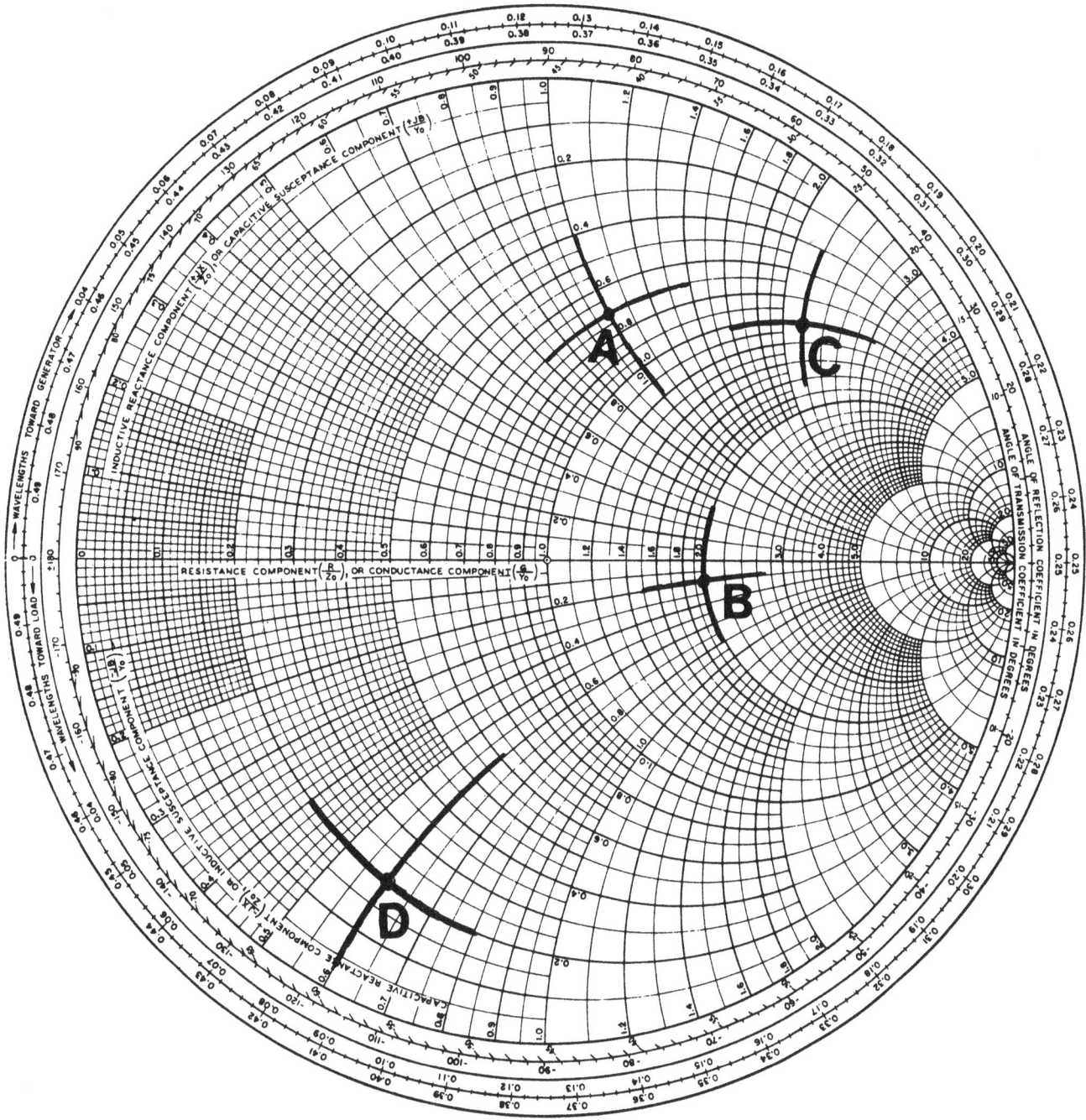

Fig. 3.2

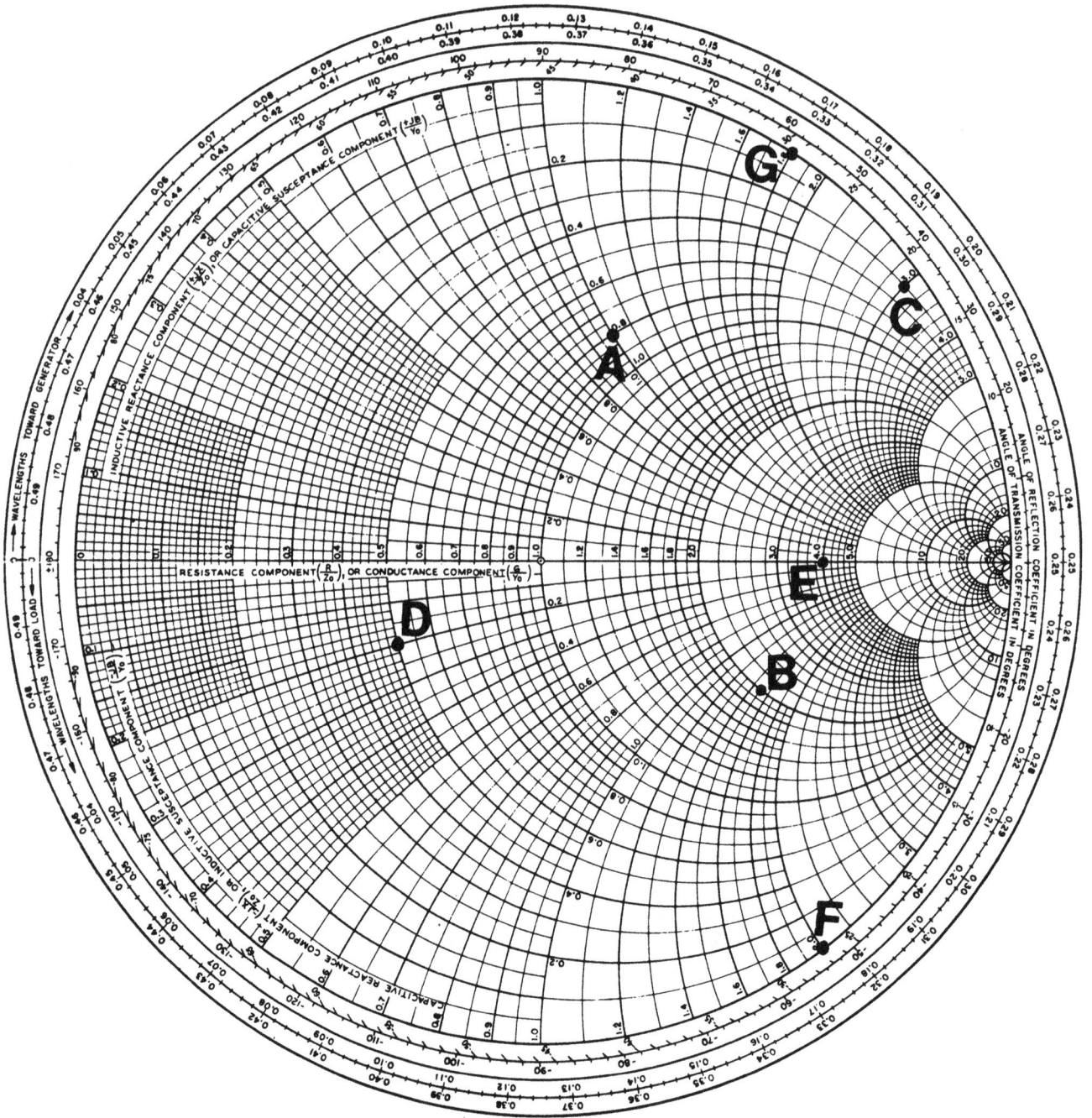

Fig. 3.3

Calculation

Impedance Z(ohm)	Normalized z_n	Point	Nature
40+j50	0.8+j1	A	R&L
100-j75	2.0-j1.5	B	R&C
10+j150	0.2+j3.0	C	R&L
25-j10	0.5-j0.2	D	R&C
200+j0	4.0+j0	E	R
0-j100	0-j2.0	F	C
0+j80	0+j1.8	G	L

Sample calculation

$Z = 40+j50$ ohm

$z_n = 40+j50$ ohm $/ 50$ ohm $= 40/50 +j50/50 = 0.8+j1$

(3.4) Read the normalized impedance values from the Smith chart in Figure (3.4). Also give the nature of the impedance.

(3.5) Write down the admittance values of the circuits at the frequencies shown in Figure (3.5A). Then normalize the admittance values against $Y_o = 0.02$ mho and plot the normalized values on a Smith chart (Figure (3.5B)).

Calculation

a. $G = 1/80$ mho $= 0.0125$ mho
$B_L = -1/2\pi \times 10^9 \times 3 \times 10^{-9}$ mho $= -0.053$ mho
$Y = 0.0125 - j0.053$ mho

$y_n = 0.0125 - j0.053$ mho $/ 0.02$ mho $= 0.625 - j2.65$ (plot as point A)

b. $G = 1/100$ mho $= 0.01$ mho
$B_C = 2\pi \times 4 \times 10^9 \times 2 \times 10^{-12}$ mho $= 0.05$ mho
$Y = 0.01 + j0.05$ mho

$y_n = 0.01 + j0.05$ mho $/ 0.02$ mho $= 0.5 + j2.5$ (plot as point B)

c. $G = 0$ (because resistance is infinite)
$B_L = -1/2\pi \times 5 \times 10^9 \times 5 \times 10^{-9} = -0.0064$ mho
$B_C = 2\pi \times 5 \times 10^9 \times 1.5 \times 10^{-12} = 0.0471$ mho
$Y = 0 + j(0.0471-0.0064)$ mho $= 0 + j0.0407$ mho

$y_n = 0 + j0.0407$ mho $/0.02$ mho $= 0 + j2.035$ (plot as point C)

d. $G = 1/200$ mho $= 0.005$ mho
$Y_L = -1/2\pi \times 4.6 \times 10^9 \times 15 \times 10^{-9}$ mho $= -0.0023$ mho
$Y_C = 2\pi \times 4.6 \times 10^9 \times 0.5 \times 10^{-12}$ mho $= 0.0145$ mho
$Y = 0.005 + j(0.0145-0.0023)$ mho $= 0.005 + j0.0122$ mho

$y_n = 0.005 + j0.0122$ mho $/ 0.02$ mho $= 0.25 + j0.61$ (plot as point D)

22

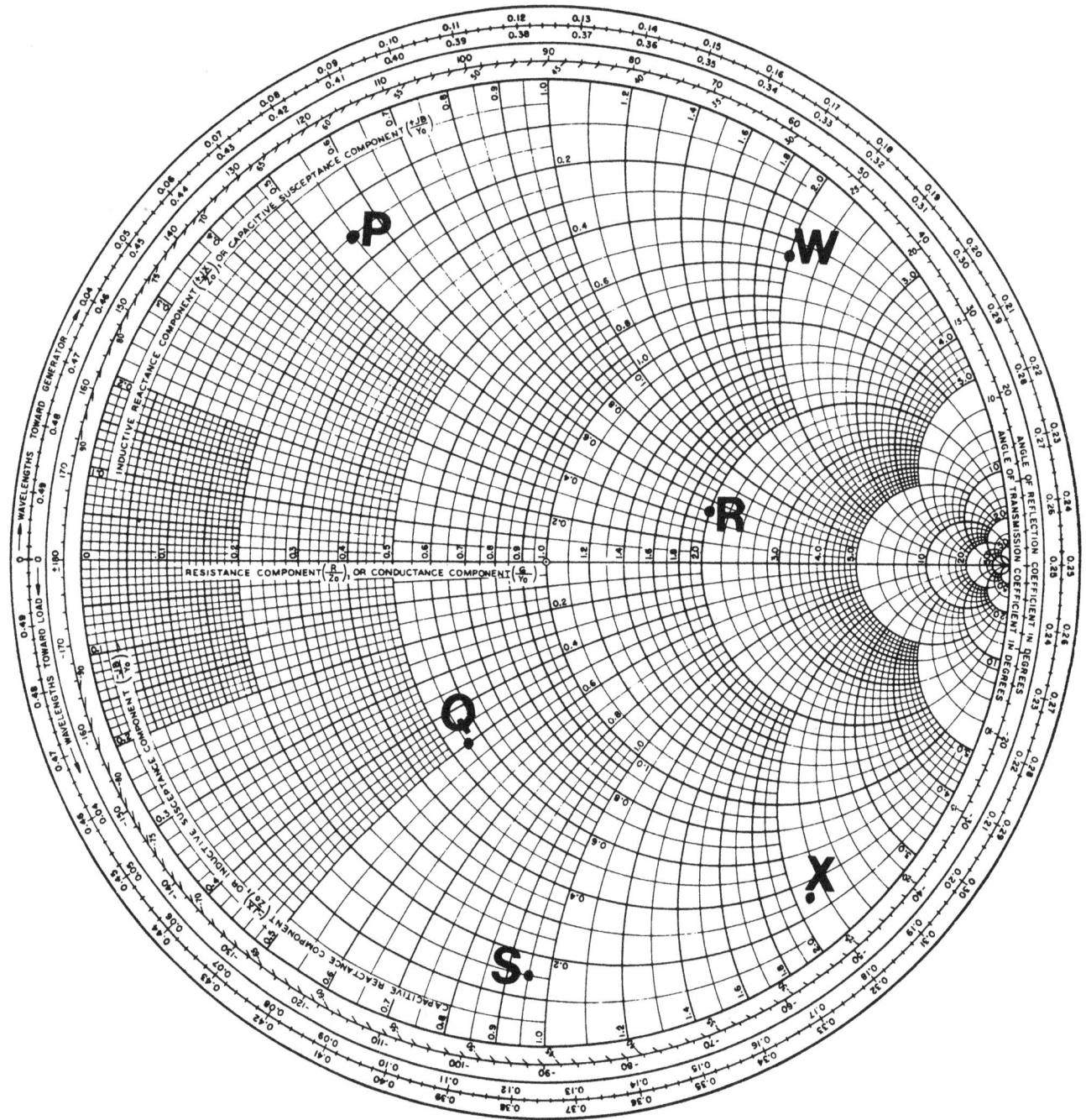

Fig. 3.4

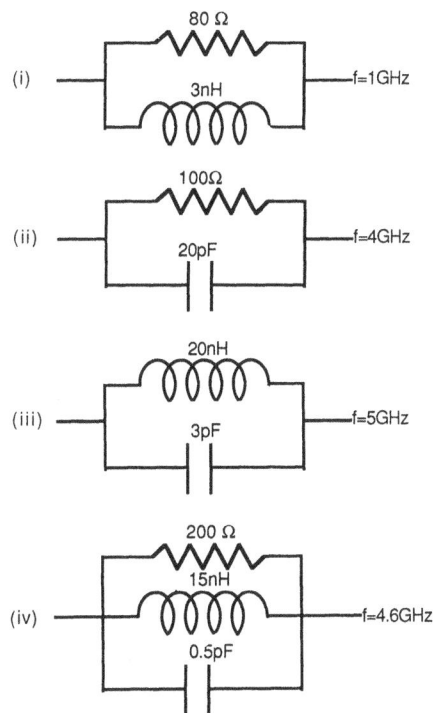

Fig. 3.5A

(3.6) Plot the following admitttance points on the Smith chart.
P: 0.6+j0.3; Q: 0.2-j0.5; R: 1-j1.6; S: 2.4+j0.2; T:3.0-j1.2.

This will be left as an exercise for the readers to verify the additional points plotted in Figure (3.5B).

(3.7) Use the Smith chart in Figure (3.6) to trace the admittance value of the normalized impedance z=0.6+j1.3.

Calculation

First plot the impedance z=0.6+j1.3, which is a resistor in series with an inductor, on the Smith chart in Figure (3.6). Draw a circle passing through z by pivoting at the center of the Smith chart. Now draw a line from z through the center of the Smith chart until it intersects the opposite side of the circle at y. The point of intersection is the normalized admittance z.

$$y = 0.3 - j0.625$$

(3.8) A load impedance $Z_L = 75+j100$ ohm is connected to a line whose impedance Z_o is 50 ohm. Use a Smith chart to find the VSWR, reflection coefficient, and return loss of the connection.

The load must first be normalized, $z_n = Z_L/Z_o = 1.5+j2.0$

The normalized impedance z_n is plotted in Figure (3.7). The VSWR, reflection coefficient (Δ), and the return loss (in dB) can be found by using the radial scale at the bottom of the Smith chart. The distance of z_n from the center of the Smith chart is measured by a compass or a ruler and then applied to the radial scale to read VSWR, Δ, and the return loss as shown.

VSWR = 4.6, Δ = 0.64   RL= 3.8dB

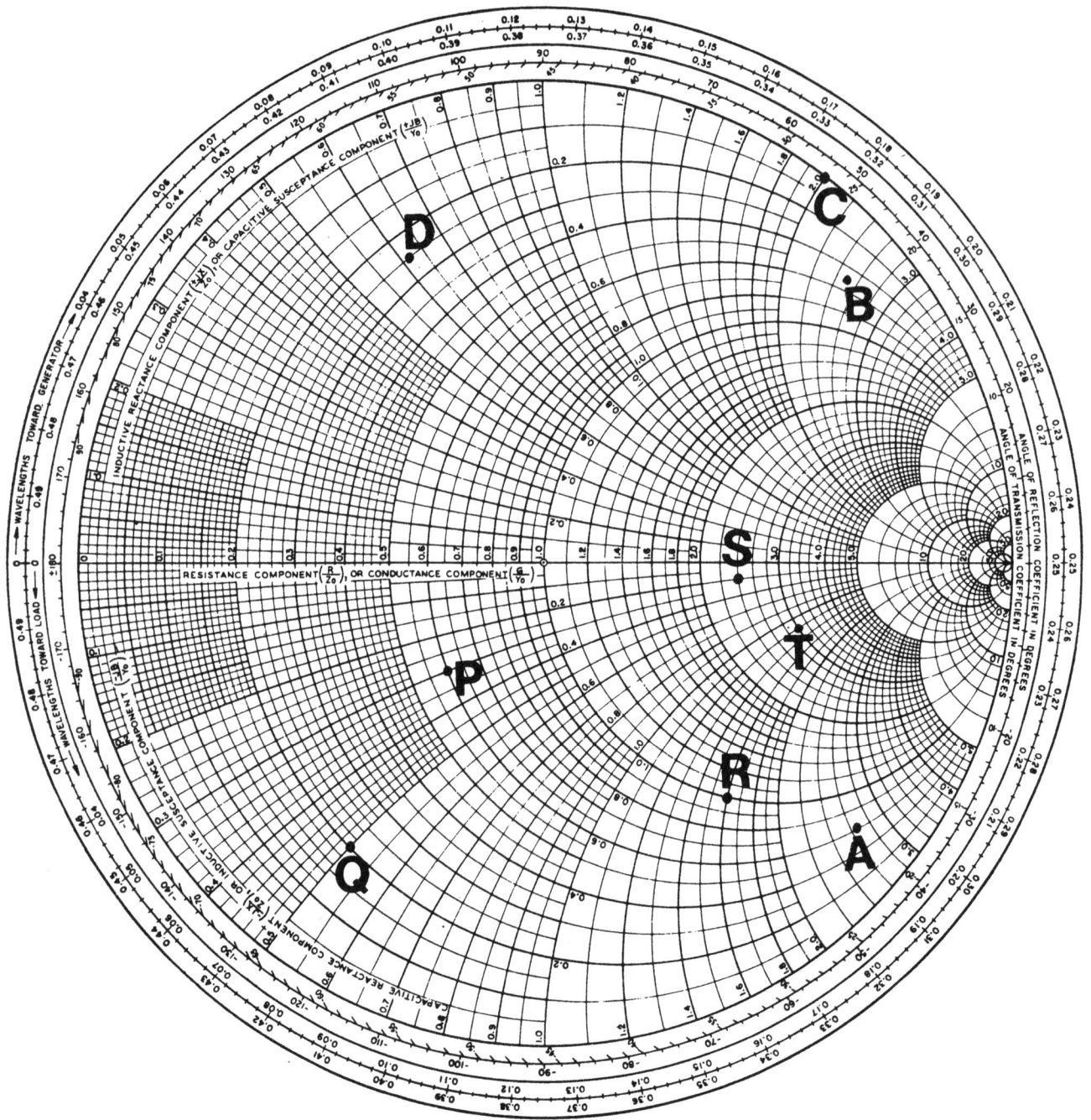

Fig. 3.5B

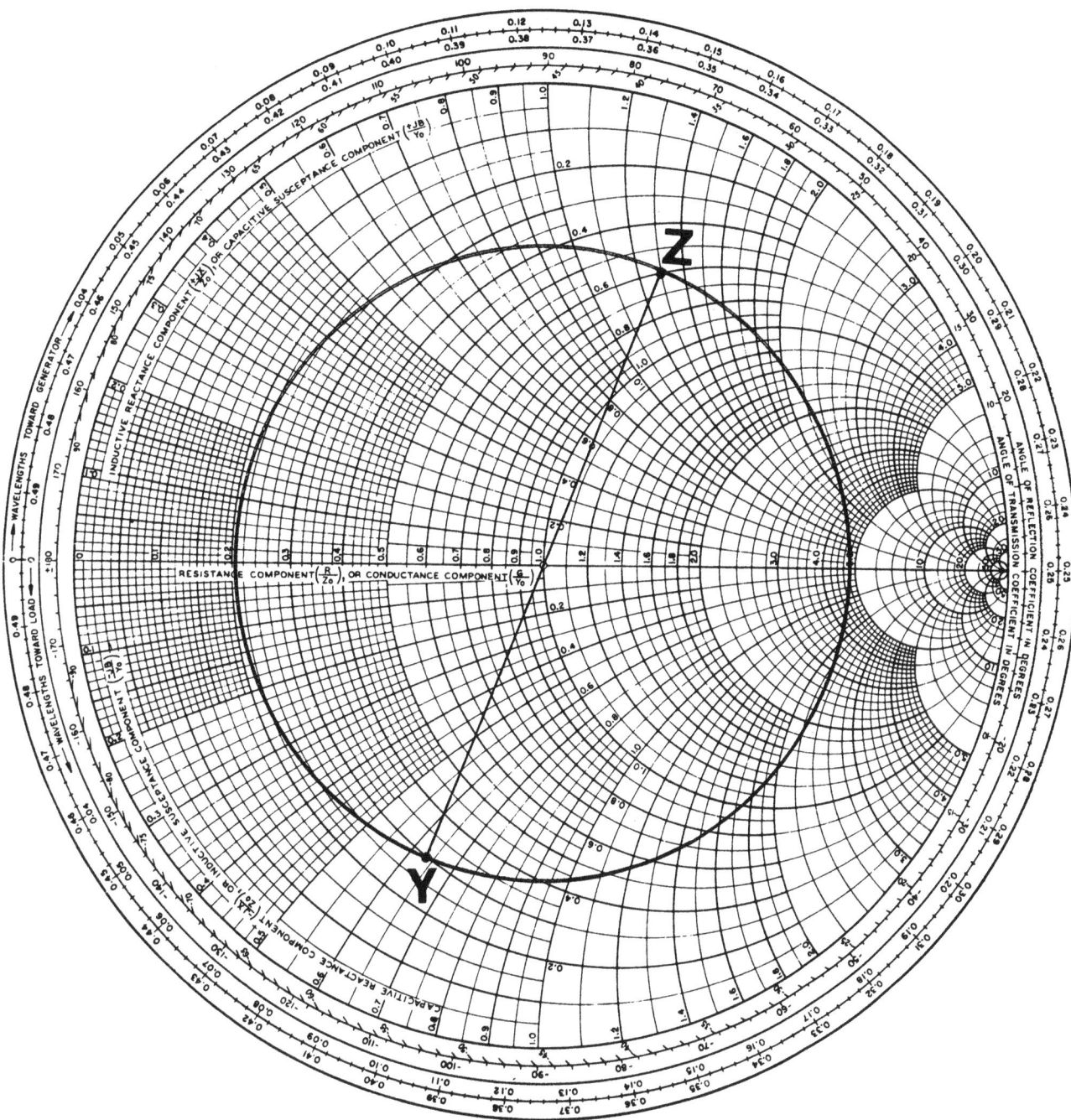

Fig. 3.6

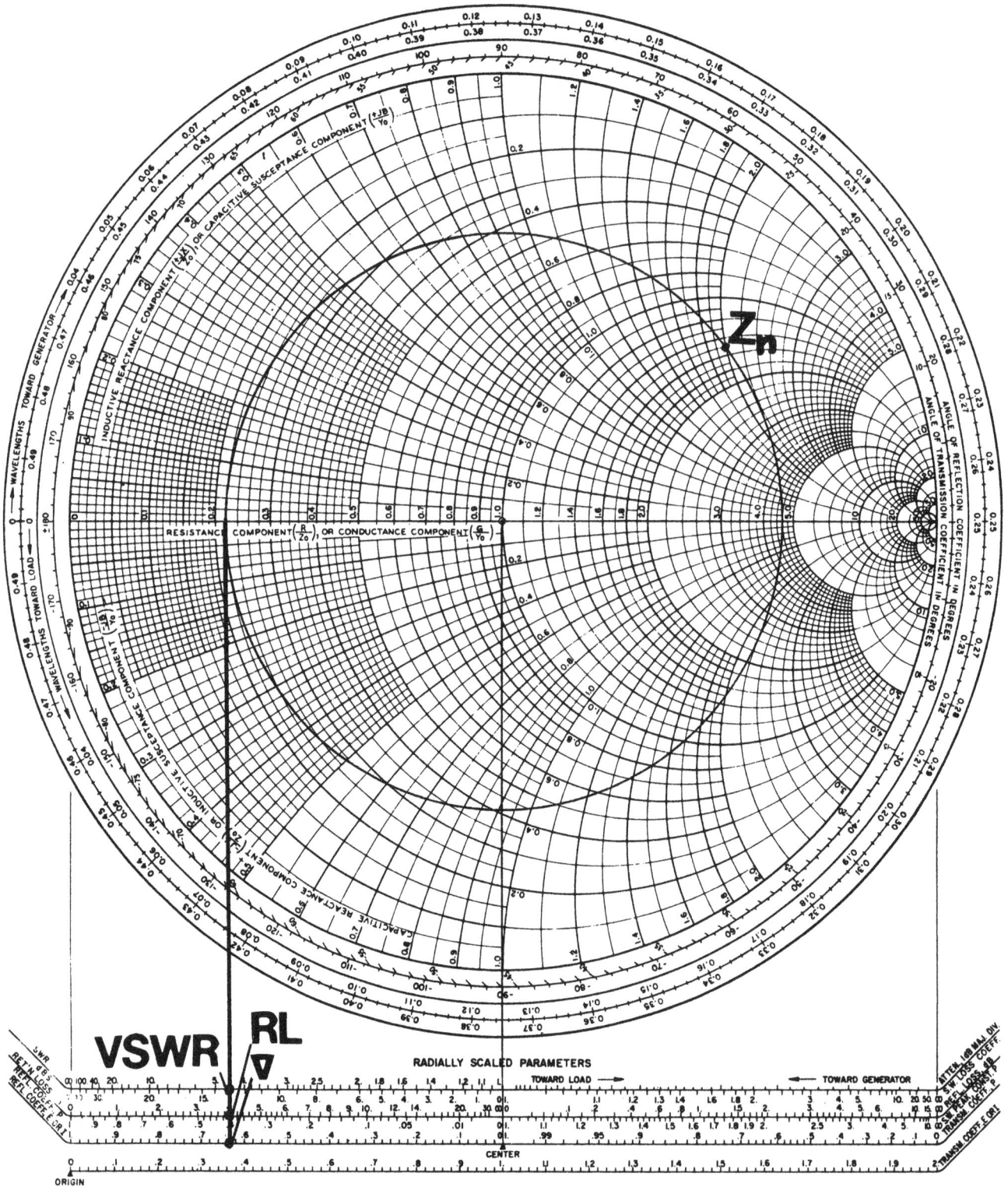

Fig. 3.7

(3.9) A load impedance Z_L=150-j225 ohm is connected to a 75 ohm line. Normalize the load impedance and then use a Smith chart to find the VSWR, reflection coefficient, and the return loss of the connection.

This will be left as an exercise for the reader.

(3.10) Find the normalized and actual impedance values at the different locations, which are 1/10 wavelength apart as shown in Figure (3.8A).

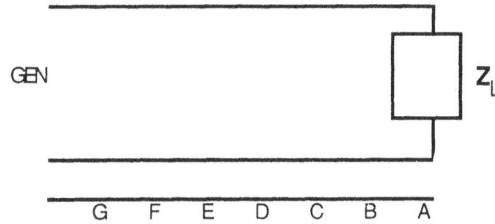

Fig. 3.8A

Calculation

The normalized load impedance is $z_n = 2 + j3$ and is plotted as point A on the Smith chart in Figure (3.8B). A circle, at the center of the Smith chart, is then drawn to pass through point A (this is called the VSWR circle). A line is drawn from the center to point A and extended to the distance movement circle in the direction of "toward generator." This is because the starting point is the load and we are moving toward the generator, i.e., clockwise.

The locations B through G are 0.1 wavelength apart. Starting from point A, which registers 0.213 on the "toward generator" circle, the locations for B through G therefore register as 0.313, 0.413, 0.513 (same as 0.013), 0.613 (same as 0.113), 0.713 (same as 0.213), and 0.813 (same as 0.313) respectively. Note that the impedance at location F is the same as that at location A and the impedance at location G is the same as that at location B.

The normalized impedances at these points are obtained as shown in Figure (3.8B) and their actual values are simply the normalized value multiplied by the line impedance 75Ω (see Exercise (3.9)).

Point	Normalized z	Actual Z (Ω)
B	1.2 -j2.4	90 - j180
C	0.22 - j0.67	16.5 - j50.25
D	0.15 + j0.02	11.25 + j1.5
E	0.23 + j0.73	17.25 + j54.75
F(same as A)	2 - j3	150 + j225
G(same as B)	1.2 - j2.4	90 - j180

(3.11) With reference to Exercise (3.10), the guide wavelength is 2 cm. Find the distance in centimeter between the load and the closest minimum of the standing wave.

Calculation

The minimum is located at the intersection between the VSWR circle and the R-axis on the left-hand side (see Figure (3.8B)). The distance is measured from the load toward the generator so the minimum register as 0.5 wavelength on the "toward generator" circle (TG circle). The distance is 0.5-0.213 = 0.287 wavelength.

28

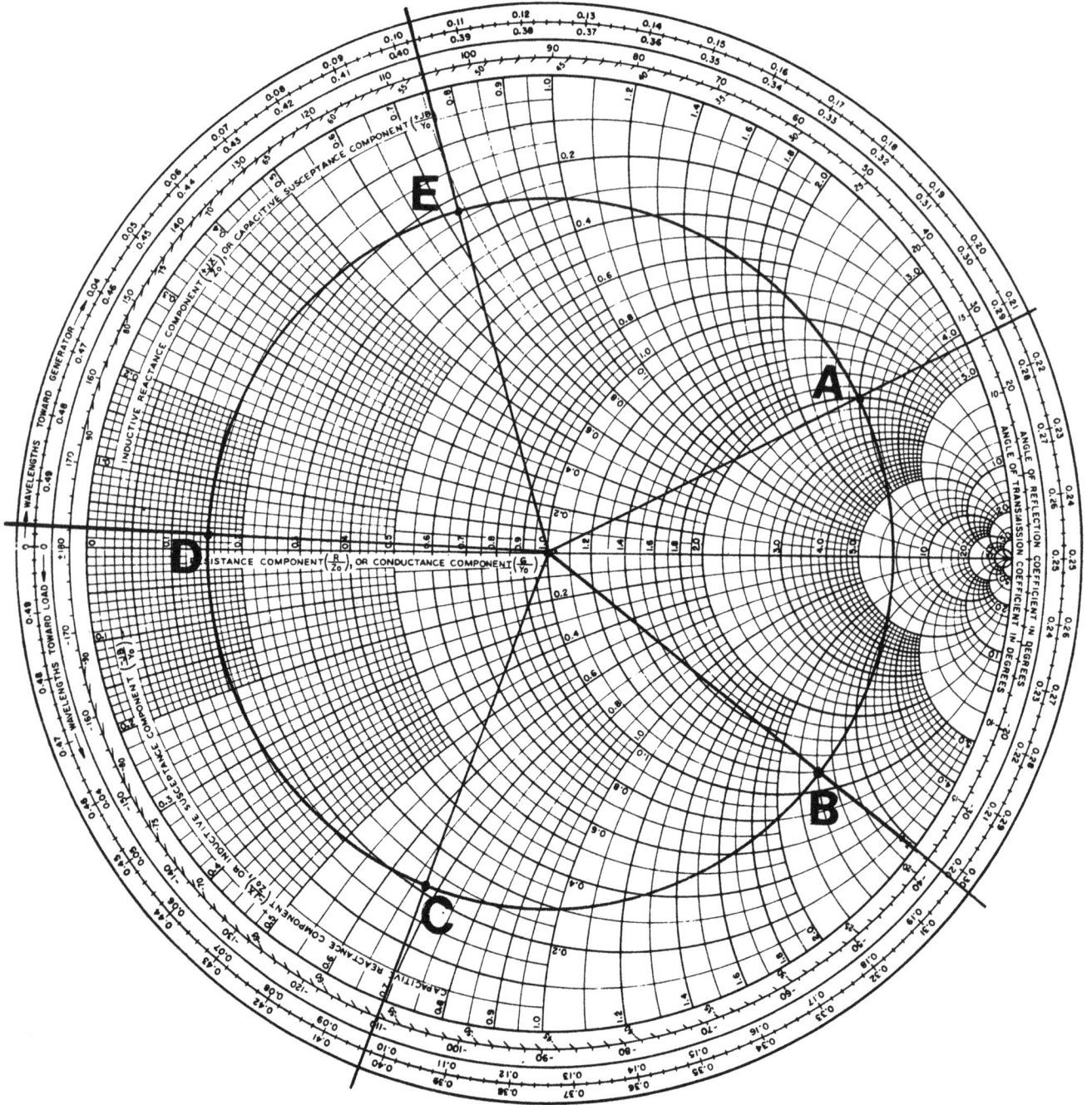

Fig. 3.8B

distance in cm = 2 cm x 0.287 =0.574 cm

(3.12) A load Z=20-j25 ohm is connected to a 50 ohm line. The guide wavelength is 2.5 cm. Find the impedance at three points, distances 0.715 cm, 1.5 cm, and 6.0 cm from the load?

The normalized load impedance is 0.4-j0.5 and is plotted as z_n in Figure (3.9).

The normalized distances of the three points are

A. 0.715 cm/2.5 cm = 0.286 wavelength
B. 1.5 cm/2.5 cm = 0.6 wavelength
C. 6.0 cm/2.5 cm = 2.4 wavelength

Starting from the load and going toward the generator (clockwise), the three points A, B, and C are located at 0.286, 0.6, and 2.4 wavelengths from the load respectively. The location of the load registers 0.417 wavelength on the "TG" circle. Point A is at 0.417+0.286 = 0.703 wavelength on the TG circle and is the same as 0.203. The impedance at point A is $z_A = 1.8 + j1.45$.

The impedances at point B and C are similarly found to be $z_B = 0.31 + j0.1$ and $z_C = 1.25 - j1.35$.

(3.13) In the previous exercise, find the admittance values of the load and the three points using the same approach.

Calculation

The admittance of the load can be found by first plotting the impedance and then rotating the chart by 180°. The admittance y_L is found to be $y_L = 0.95 + j1.25$ as shown in Figure (3.10).

Starting from y_L and going toward the generator by 0.286, 0.6, and 2.4 wavelengths as usual, the normalized admittance values of the three points are found to be

$$y_A= 0.32 - j0.27, \qquad y_B= 0.29 - j1.0, \qquad y_C= 0.35 + j0.41$$

(3.14) A load 20+j30 ohm is connected to one end of a 50 ohm line which is 4 cm long. The other end of the 50 ohm line is connected to a 75 ohm line as shown in Figure (3.11A). The guide wavelength is 2.5 cm. Find the impedance at point P which is 10 cm from the load.

Calculation

The load impedance normalized against 50 ohm is z_L=0.4+j0.6 and is plotted on the Smith chart as shown in Figure (3.11B).

The distance between the load and the joint between the 50 ohm and the 75 ohm line is given to be 4 cm. The normalized value of this length is 4cm/2.5cm =1.6 wavelength. The load registers 0.094 on the TG cirlce. Starting from the load and going toward the generator by a distance of 1.6 wavelength, the normalized impedance at the joint is found to be $z_{J1} = 1.5 + j1.55$.

The actual impedance at the joint is Z_{J1}:

$$Z_{J1}= 50 \text{ ohm} \times z_{J1} = 75 + j77.5 \text{ ohm}$$

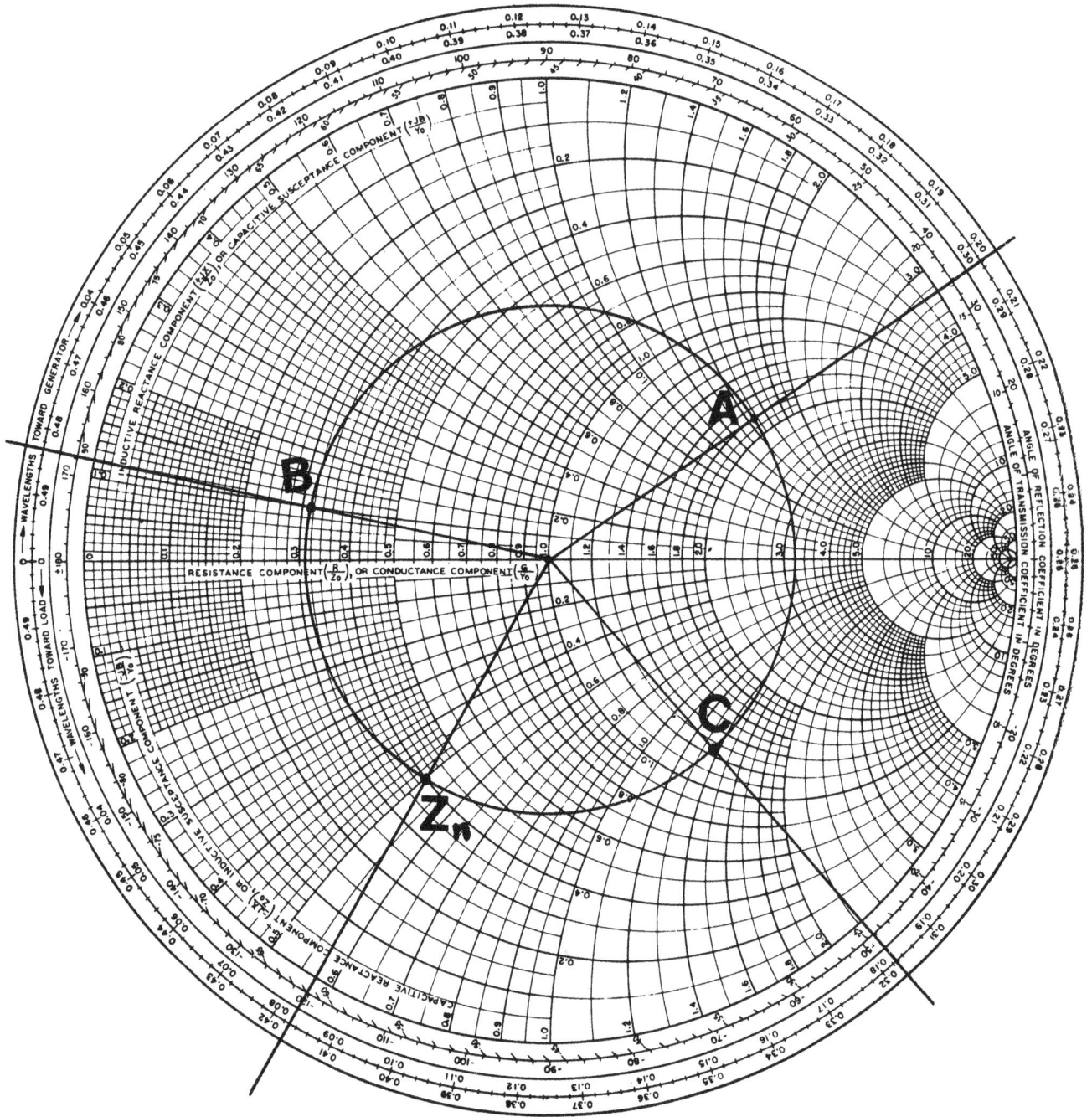

Fig. 3.9

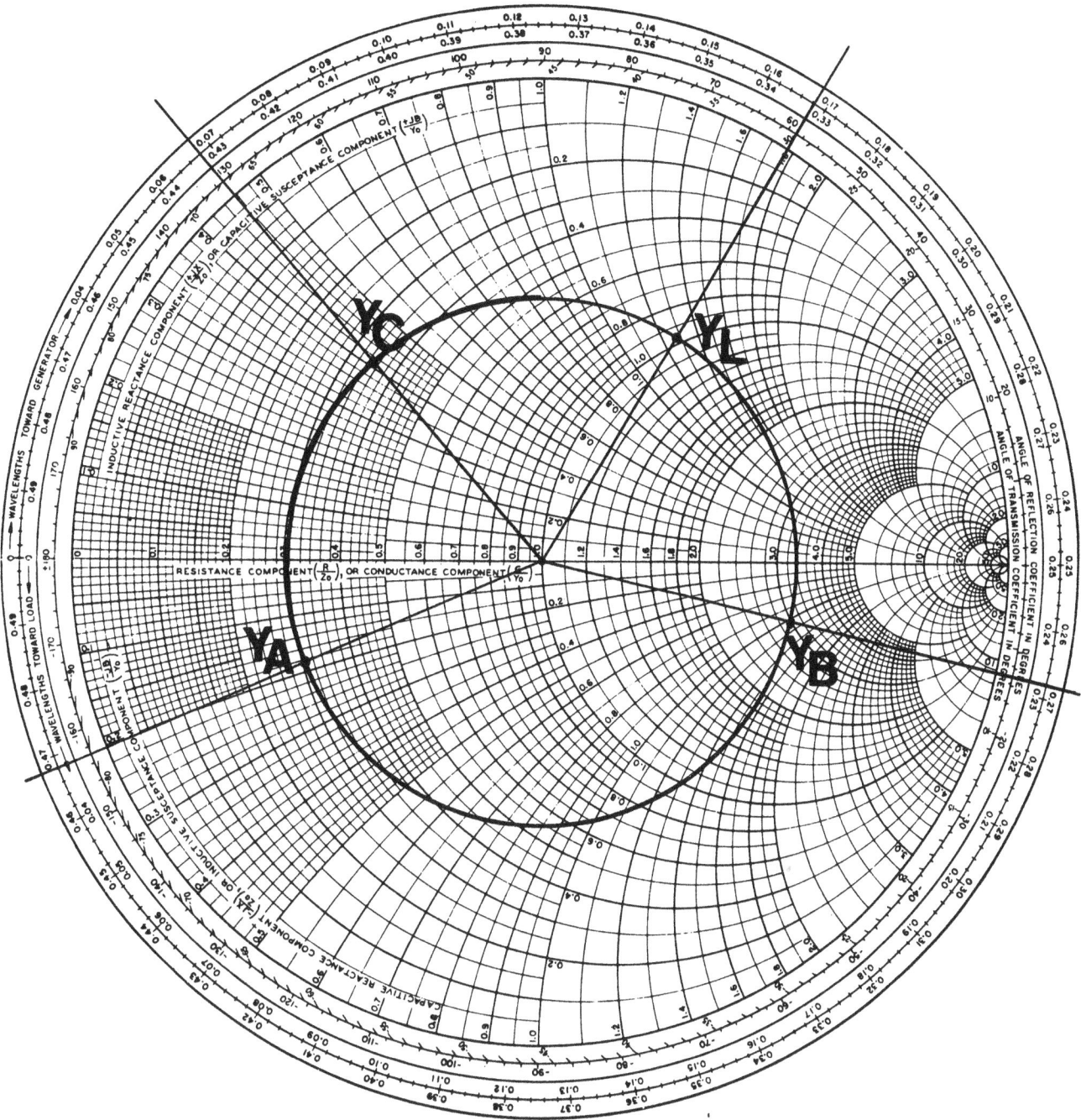

Fig. 3.10

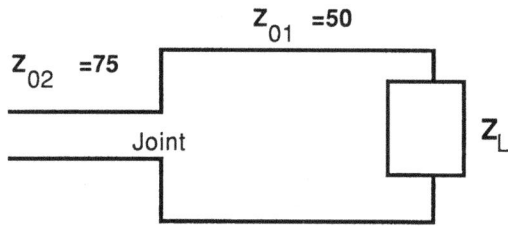

Fig. 3.11A

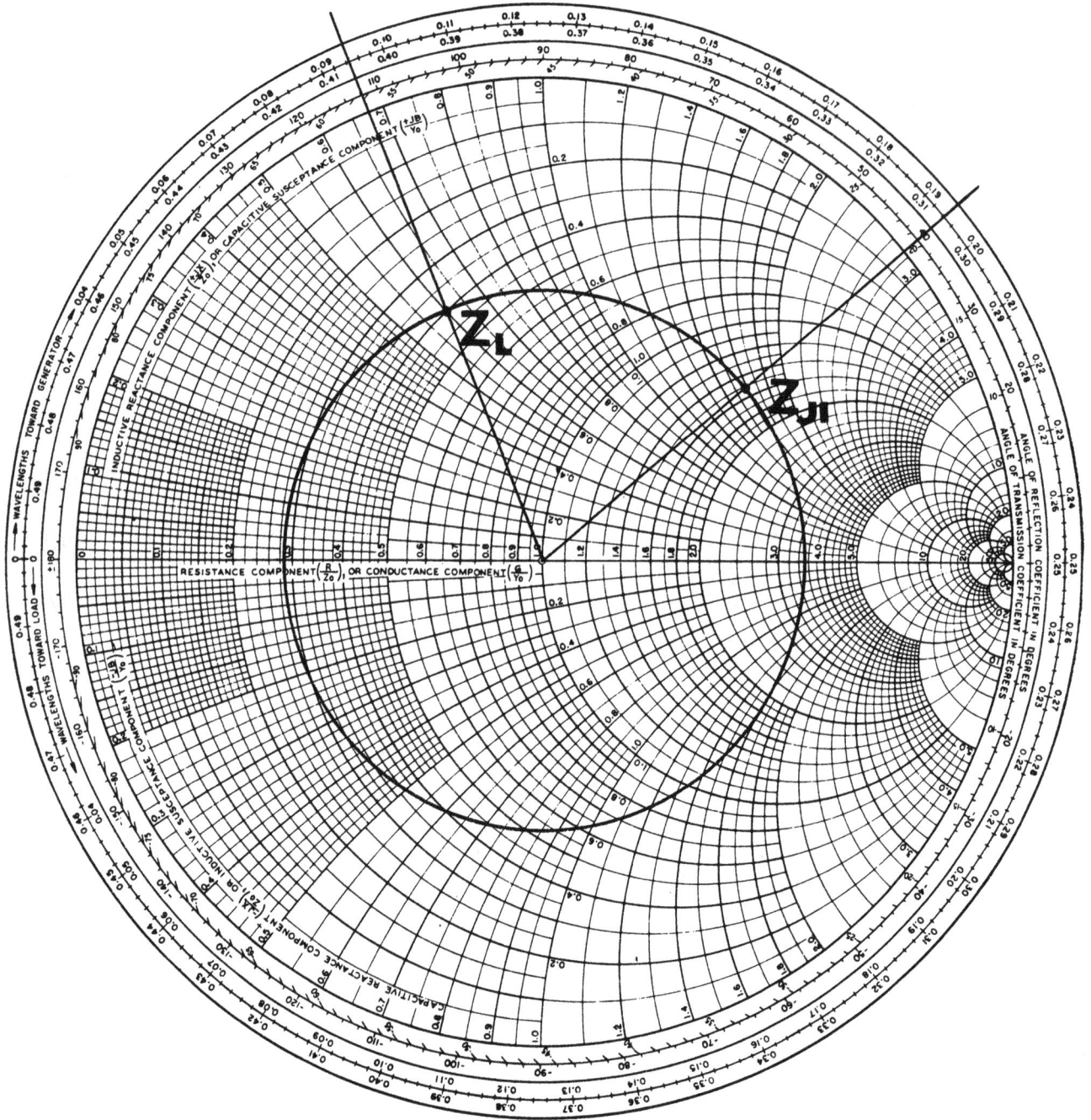

Fig. 3.11B

We then normalize Z_{J1} against the 75 ohm line and call it z_{J2}:

$$z_{J2} = (75 + j77.5 \text{ ohm})/75 \text{ ohm} = 1.0 + j1.033$$

The normalized impedance z_{J2} is plotted on another Smith chart (Fig.3.11C). The distance between point P and the joint is 10cm - 4cm = 6cm. The normalized distance is 6cm /2.5 cm = 2.4 wavelengths.

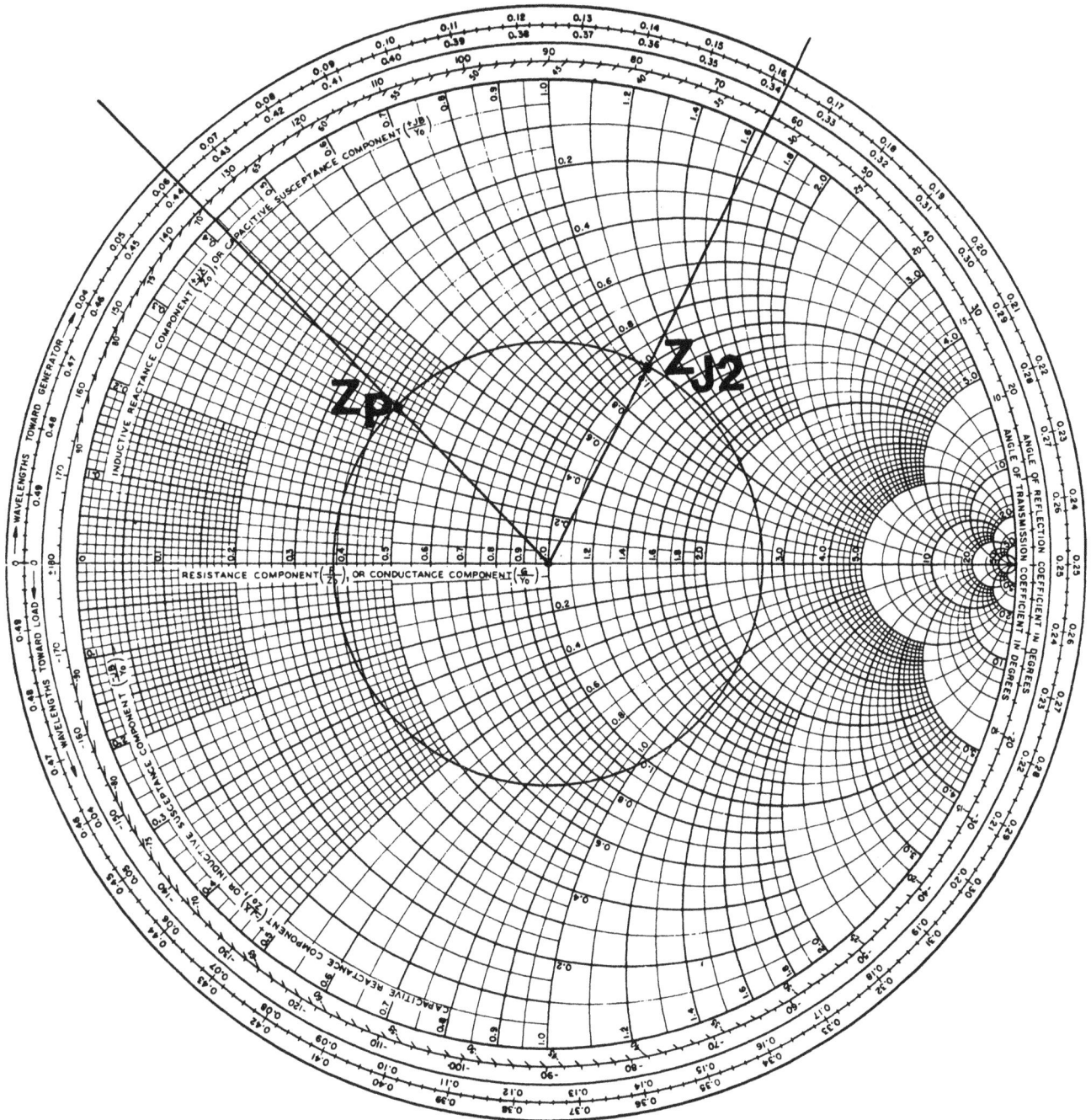

Fig. 3.11C

Starting from z_{J2} and going toward the generator by a distance of 2.4 wavelength, we find the normalized impedance of point P to be $z_P = 0.42 + j0.35$.

The actual value of Z_P is 75 ohm x (0.2 + j0.35) = 15 + j26.25 ohm.

(3.15) Explore the principle of matching of a resistive load using a quarter-wave transformer.

Calculation

Consider a purely resistive load R_L ohm connected to a line (known as the quarter wave transformer) whose length is 1/4 the guide wavelength as shown in Figure (3.12). The line impedance of the quarter-wave transformer is chosen to be Z_T (by adjusting the line width relative to the normal Z_0 in the rest of the circuit) so the normalized load is R_L/Z_T. The (normalized) impedance seen at the left of the quarter-wave line can be found easily by rotating the Smith chart by 180° as illustrated in Exercise (3.7). We will call this (normalized) impedance z_{in} for the input impedance looking to the right of the quarter-wave line.

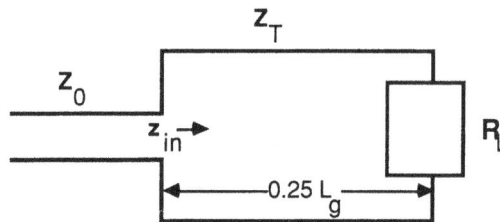

Fig. 3.12

Mathematically, z_{in} is simply the reciprocal of the normalized load impedance, i.e., $z_{in} = Z_T/R_L$. The actual impedance of z_{in} is $Z_{in} = z_{in} \times Z_T = Z_T^2/R_L$ ohm.

For matching, the input impedance Z_{in} must appear to the preceeding circuit to be equal to the ordinary line impedance Z_0. In other words, $Z_T^2/R_L = Z_0$.

This gives a very useful formula, $Z_T = \sqrt{Z_0 R_L}$ ohm. If expressed in a normalized form, $z_T = Z_T/Z_0 = \sqrt{R_L/Z_0}$, i.e. the square root of the normalized value of the load resistance.

For example, a resistive load of 70 ohm is to be matched to a 50 ohm circuit by means of a quarter wave transformer, the transformer impedance Z_T is calculated according to the preceeding formula to be

$$Z_T = \sqrt{50 \times 70} \text{ ohm } = 59.16 \text{ ohm.}$$

The normalized value of Z_T is either $z_T = 59.16$ ohm/50 ohm = 1.18, or using the second equation to get $z_T = \sqrt{70/50} = 1.18$.

(3.16) What is the principle of matching of an ordinary load impedance by means of a quarter wave transformer?

Calculation

The load impedance must first be transformed to become purely resistive by adding a small line length of either s_1 or s_2. Length s_1 will bring the effective impedance to simple resistance (with no reactance) to the left, and s_2 to the right, of the center of the Smith chart. Now that the effective impedance is purely resistive, the method described in the previous question can be applied.

For example, the guide wavelength of a circuit is 2 cm and the line impedance of the circuit is 50 ohm. A load of normalized impedance 0.8-j1.4 (as shown in Figure (3.13)) is to be matched by a quarter wave transformer. Two line lengths are possible. The first length s_1 will bring z_L to the left of the Smith chart resulting in an effective load impedance (normalized) equal to $z_1 = 0.24 + j0$ and the normalized length of s_1 equal to 0.5 - 0.334 = 0.166 wavelength. The actual length is 0.166x2cm = 0.332cm. The normalized transformer impedance z_T is then the square root of the normalized value of the effective load impedance, i.e., $z_T = \sqrt{0.24} = 0.49$.

The second length s_2 will bring z_L to the right of the Smith chart resulting in an effective load impedance (normalized) equal to $z_2 = 4.2 + j0$ and the normalized length of s_2 equal to 0.166+0.25 = 0.416. The actual length is 0.416x 2cm = 0.832cm. The normalized transformer impedance z_T is then equal to $\sqrt{4.2} = 2.05$.

(3.17) A load 70 + j100 ohm is attached to a 50 ohm line. Describe the method of matching by adding a series component at the appropriate location of the line. The frequency of the circuit is 5GHz and the guide wavelength is 4 cm.

Calculation

The normalized load impedance is 1.4 + j2.0 and is plotted on a Smith chart as shown in Figure (3.14). The idea of matching is to describe a locus that would bring the effective load impedance to the center of the Smith chart, i.e., the overall impedance is 1.0 + j0.

When moving from the load toward the generator, the impedance follows the VSWR circle as shown. With reference to Figure (3.14), location A is desirable because the resistance part at A is 1.0. Actually, the impedance at A is 1.0 - j1.7, which means it behaves electronically like a resistor in series with a capacitor, the reactance of which is -j1.7. By adding a series inductor of reactance +j1.7, the overall impedance is now 1.0 + j0, i.e., matched. The actual value of the inductive reactance is 1.7x50 ohm =85 ohm. The inductance L of this component is found by solving the equation

$$2\pi fL = X_L = 85$$

which gives $L = 2.7 \times 10^{-9}$ H = 2.7 nH. The distance of travel from the load to point A is 0.318 - 0.196 = 0.122 wavelength. The actual distance is 0.122x 4 cm = 0.488 cm.

An alternative location is to pass point A and proceed to point B whose resistance part is also 1.0. The impedance at point B is 1.0 + j1.7, which means it behaves electronically like a resistor in series with an inductor. By adding a series capacitor of reactance -j1.7, the overall impedance now becomes 1.0 + j0. The actual value of the capacitive reactance is 1.7x50 ohm = 85 ohm. The capacitance of this component is found by solving the equation

36

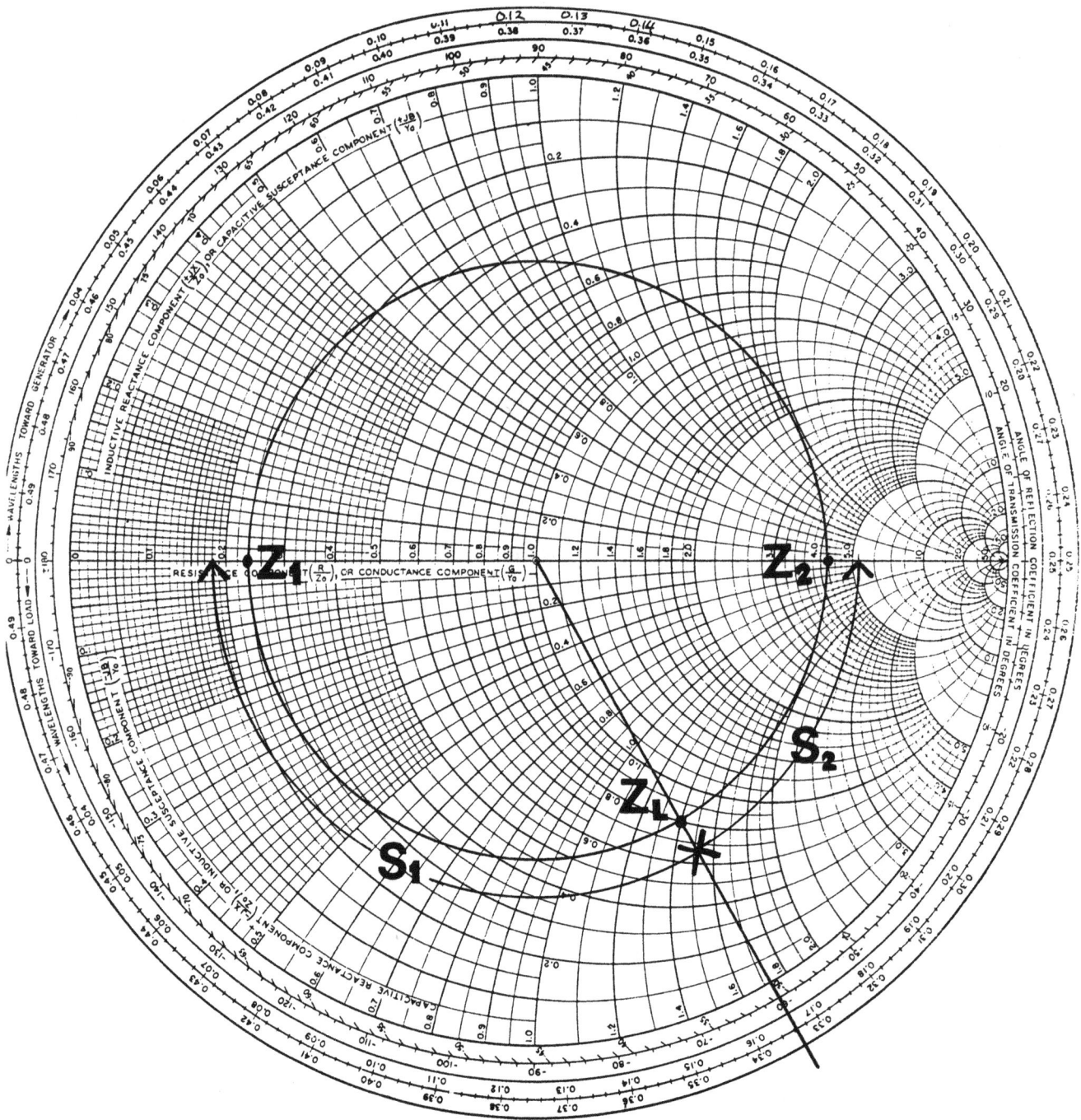

Fig. 3.13

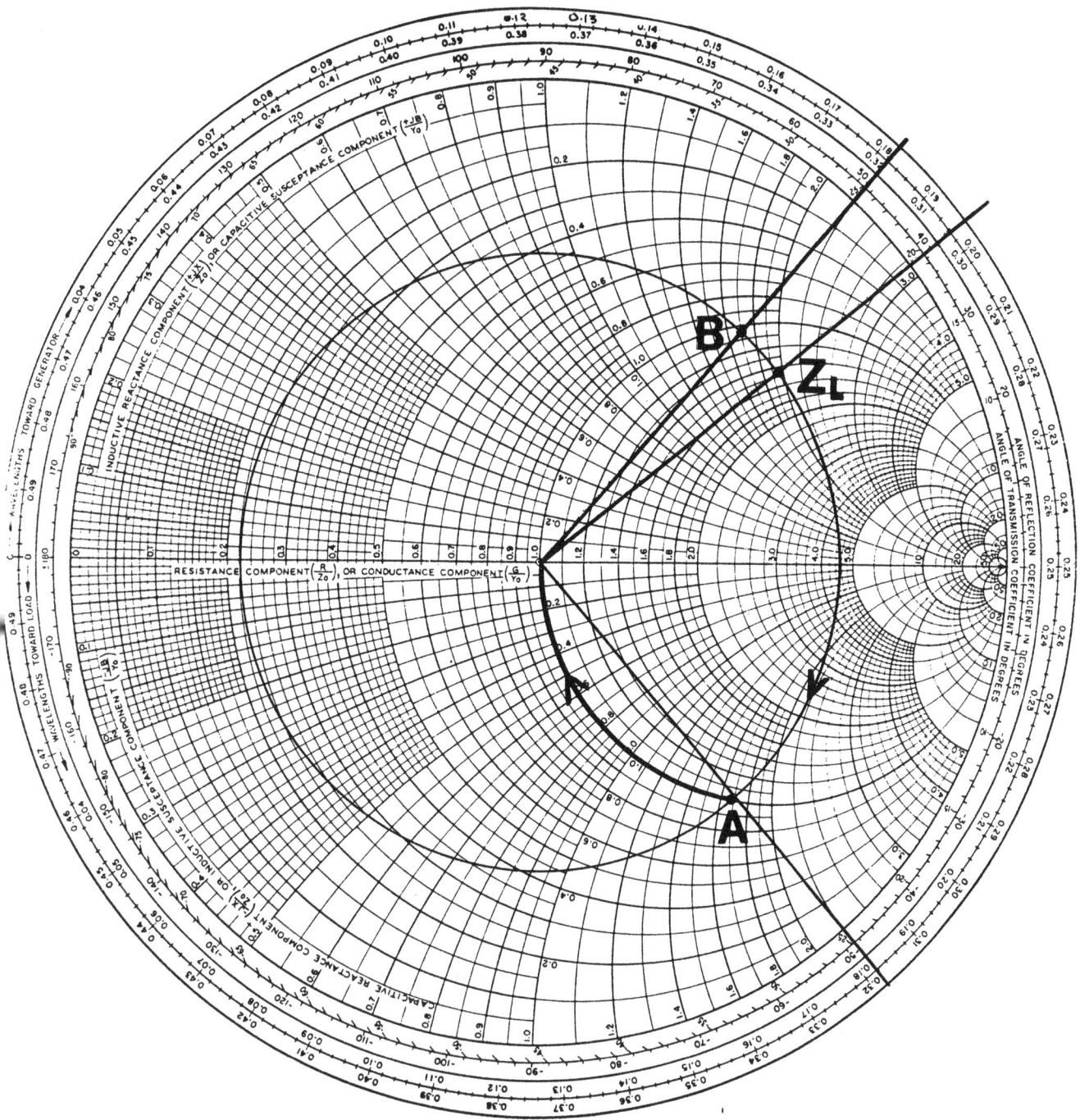

Fig. 3.14

$$1/2\pi fC = X_C = 85$$

which gives $C = 3.74 \times 10^{-13}$ F = 0.34 pF. The distance of travel from the load to point B is 0.5 + 0.182 - 0.196 = 0.486 wavelength. The actual distance is 0.486x4 cm = 1.944 cm.

Therefore, the circuit can be matched by adding in series either a 2.7 nH inductor at 0.488 cm from the load or a 0.34 capacitor at 1.944 cm from the load.

(3.18) Match the previous example by parallel components. The frequency and the guide wavelength remain the same.

Solution:

Parallel matching requires switching to the admittance coordinates. Therefore, the admittance of the load must be obtained. The reader can verify that the admittance of z= 1.4 + j2.0 is y = 0.24 - j0.34 and is plotted on a Smith chart as shown in Figure (3.15). It must be remembered that we are now dealing with admittances. The admittance of the 50 ohm line is 1/50 = 0.02 mho.

Starting from the load and moving toward the generator along the VSWR circle, location C is desirable because the conductance part of the admittance is 1.0. The admittance at C is 1.0 + j1.7, which is equivalent to a resistor in parallel with a capacitor. By adding an inductor in parallel at C, the overall admittance can be brought to 1.0 + j0. The normalized susceptance of the needed inductor is -j1.7 and the actual value is 1.7 x 0.02 mho = 0.014 mho. The inductance of the inductor is found by solving the equation

$$1/2\pi fL = 0.014$$

which gives $L = 2.27 \times 10^{-9}$ H = 2.27 nH. The distance of travel is 0.5 + 0.18 - 0.446 = 0.234 wavelength. The actual distance is 0.234 x 4 cm = 0.936 cm from the load.

An alternative is again to pass point C and proceed to point D where the conductance is also 1.0. The admittance at point D is 1.0 - j1.7, which is equivalent to a resistor in parallel with an inductor. By adding a capacitor in parallel at D, the overall admittance becomes 1.0 + j0. The susceptance of the capacitor is +j1.7 and the actual value is 0.014 mho. The capacitance of the capacitor is found by solving the equation

$$2\pi fC = 0.014$$

which gives $C = 4.46 \times 10^{-13}$ F = 0.446 pF. The distance of travel is 0.5 + 0.32 - 0.446 = 0.374 wavelength. The actual distance is 0.374 x 4 cm = 1.496 cm from the load.

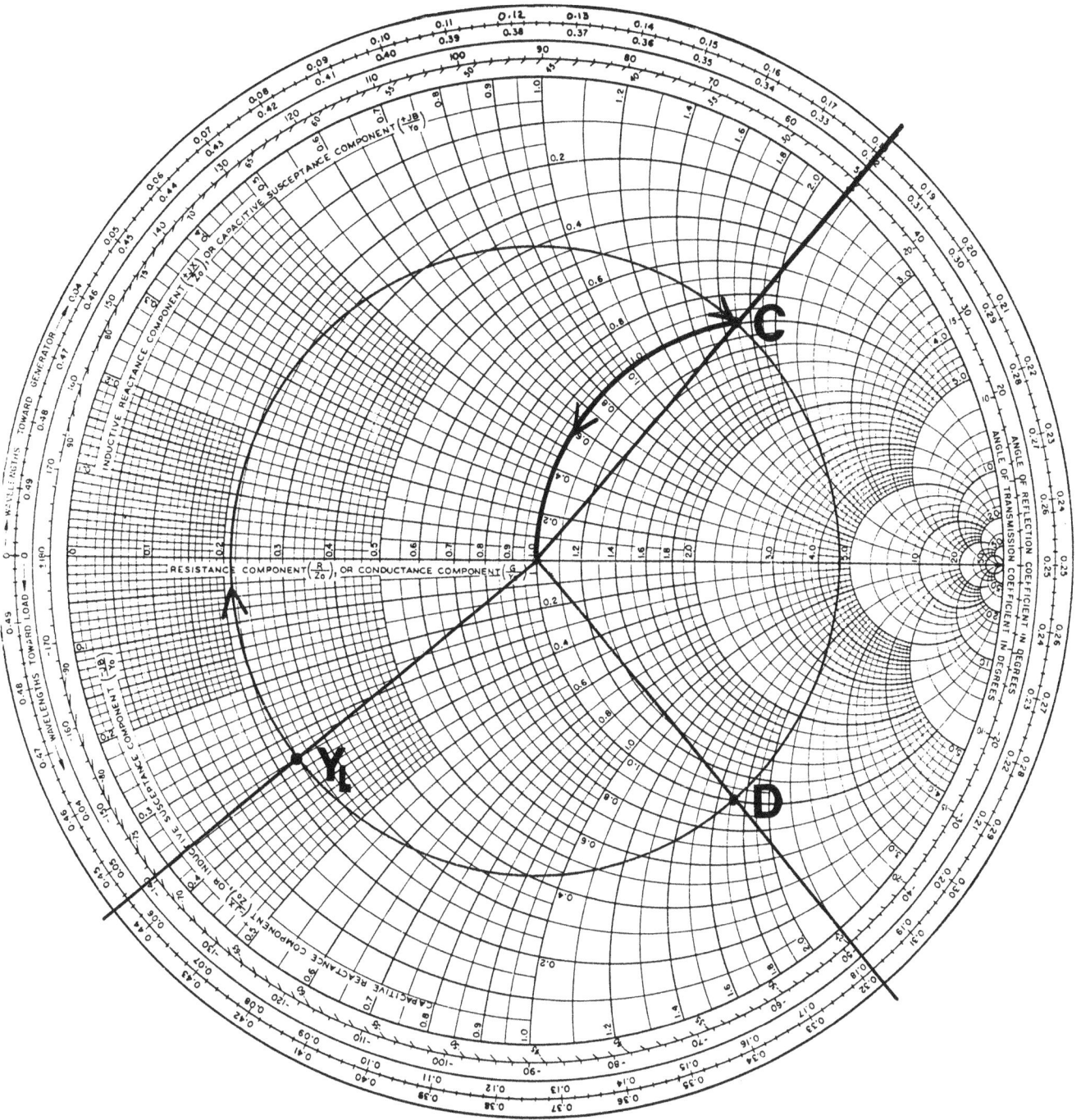

Fig. 3.15

4

SOLID-STATE AMPLIFIERS AND OSCILLATORS

4.1 BIPOLAR TRANSISTOR AMPLIFIERS

4.1.1 Ohm's Law

Using Ohm's law and both Kirchhoff's voltage and current laws, it becomes possible to solve most dc circuit problems. Ohm's law states that the current I through a linear resistor R is equal to the voltage V across it divided by the resistance. Note that the direction of current flow is from a higher potential to a lower potential.

$$I = V/R \qquad\qquad (4.1)$$

Exercises

(4.1) For the circuit in Figure (4.1), find the current and indicate its direction with an arrow.

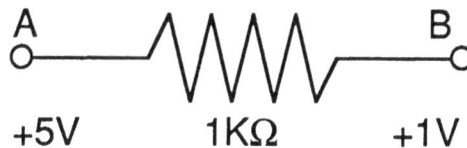

Fig. 4.1

Calculation

Begin by finding the voltage across the 1kΩ resistor. The voltage across the resistor is defined as the voltage drop V from point A to point B.

$$V_{AB} = V_A - V_B = 5V - 1V = 4V.$$

The current I, from A to B, is

$$I = V_{AB} /1k\Omega = 4V/1k\Omega = 4mA.$$

(4.2) Use Figure (4.2) to find the values of current and its direction from the given voltages and resistors.

Voltage at A	Resistance(R)	Current(I)	Direction
10V	500Ω		
5V	100Ω		
7V	700Ω		
5V	1kΩ		
12V	200Ω		

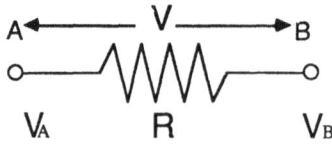

Fig. 4.2

(4.3) Use Figure (4.3) to find the values of current and its direction from the given voltages and resistors.

V_A	V_B	R	I	Direction
12V	5V	1kΩ		
10V	8V	100Ω		
5V	10V	1kΩ		
2V	8V	200Ω		
1V	10V	450Ω		

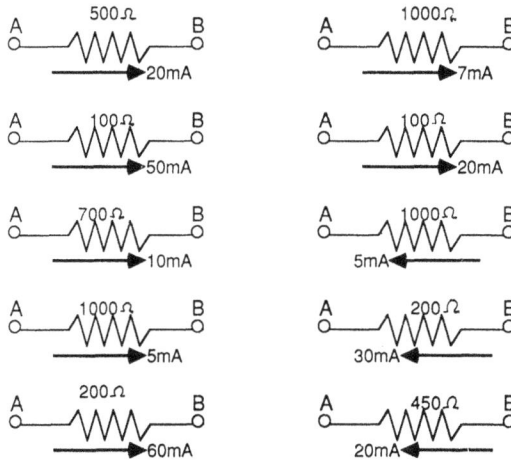

Fig. 4.3

4.1.2. Kirchhoff's Voltage Law

The sum of the voltage drops around any closed current path loop is equal to zero volt, 0V.

(4.4) For the circuit in Figure (4.4), find the current I if V_{CC} is 10V and R_1 is 800Ω and R_2 is 200Ω.

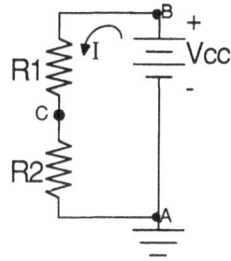

Fig. 4.4

Calculation

First define points A, B, and C. Now follow a closed current path loop beginning at A, through points A and C, and back to A. Remembering the rule, the first voltage drop from A to B is actualy a voltage rise. A voltage rise is opposite to a voltage drop and is indicated by a negative sign. Therefore the first term in the equation is $-V_{CC}$. The next voltage drop from B to C is the current I multiplied by R_1 via Ohm's law. We are now at point C in our path. Going from C to A, another voltage drops occurs which is IR_2. Since we are back where we started at point A, this equation equals 0V.

$$-V_{cc} + IR_1 + IR_2 = 0$$

Substituting, we have

$$-10V + I(800) + I(200) = 0$$
$$I = 10mA$$

(4.5) Find the current I for the following voltages and resistances as shown in Figure (4.5).

V_{CC}	R_1	R_2	I
8V	1kΩ	7kΩ	
10V	100Ω	900Ω	
6V	150Ω	50Ω	
12V	200Ω	1kΩ	
11V	500Ω	50Ω	

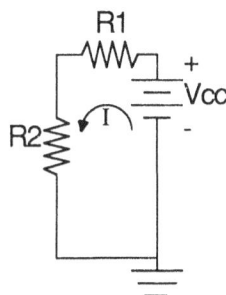

Fig. 4.5

(4.6) Find the current I for the following voltages and resistances as shown in Figure (4.6).

V_{CC}	R_1	V_{BB}	I
12V	1kΩ	5V	
15V	200Ω	7V	
6V	1kΩ	10V	
2V	500Ω	12V	
10V	400Ω	0.8V	
4V	400Ω	12V	

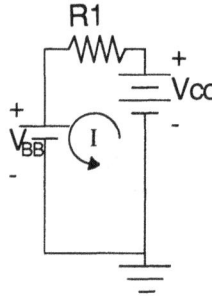

Fig. 4.6

4.1.3. Kirchhoff's Current Law

The sum of all the currents entering a node or junction point is equal to the sum of the all the currents leaving a node or junction. A node or junction can be simply thought of as a water pipe junction with water flowing in and out. All the water flowing in must be equal to all the water flowing out.

For example, Figure (4.7) shows that the sum of all the currents entering the node is I_1. The sum of all the currents leaving is I_2 and I_3. Therefore, $I_1 = I_2 + I_3$. If I_1 is 10mA and I_2 is 8mA, then I_3 is found to be 2mA.

(4.7) Use Figure (4.8) to calculate the unknown current in each of the following problems using Kirchhoff's current law.

I_1(mA)	I_2(mA)	I_3(mA)	I_4(mA)
10	5		7
20	2	10	
8		6	7
	50	32	24
10	6	0	
8	0		3
0	16	7	
12	8		15
42		15	30
	5	21	20

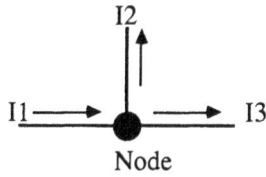

Fig. 4.7

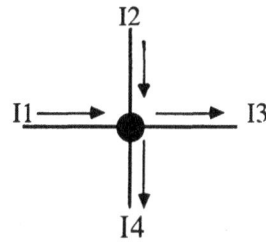

Fig. 4.8

4.1.4. Resistor Biasing Using One Battery

To analyze the following types of problems, we must apply Ohm's law, Kirchhoff's voltage and current laws. We also need to use the transistor I/V chart to obtain the relationship of base and collector currents. See Section 11.21 in the reference text. For the following illustration, the I /V chart in Figure (4.9) will be used.

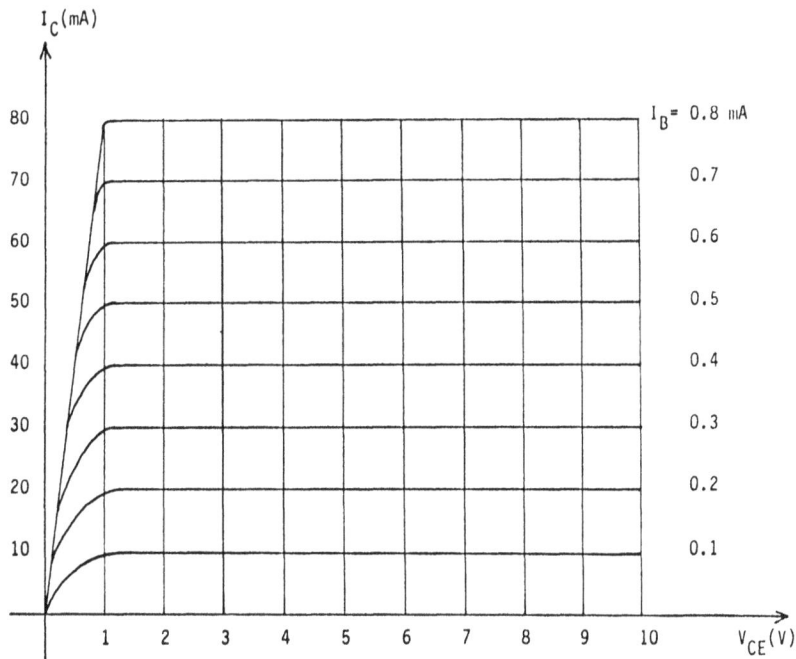

Fig. 4.9

(4.8) Given V_{CC} =12V, V_{BE} =1V, I_B =0.2mA, I_C = $9I_B$, R_C =500Ω, find I_C, V_{CE}, R_1, and R_2 as shown in Figure (4.10A).

Calculation

Step 1- Using the I /V chart in Figure (4.9), the collector current I_C can be predicted given the base current I_B. If I_B = 0.2mA, then I_C = 20mA.

Step 2- Isolate the collector-emitter circuit as shown in Figure (4.10B). Also, we need to define the voltage drop across collector to emitter of the transistor V_{CE}. Note that a closed current path can be seen. Apply Kirchhoff's voltage law by starting at point G.,

$$-V_{CC} + I_C R_C + V_{CE} = 0$$

Substituting known quantities, we have

$$-12V + (20mA)(500\Omega) + V_{CE} = 0$$
$$V_{CE} = 2V$$

Step 3- Isolate the base-emitter circuit as shown in Figure (4.10C). Note that point A is a current node. Applying Kirchhoff's current law, we have $I_1 = I_B + I_2$.

Since $I_2 = 9I_B = 9(0.2mA) = 1.8mA$, therefore $I_1 = 0.2mA + 1.8mA = 2.0mA$.

By applying Ohm's Law, we calculate the values of R_1 and R_2:

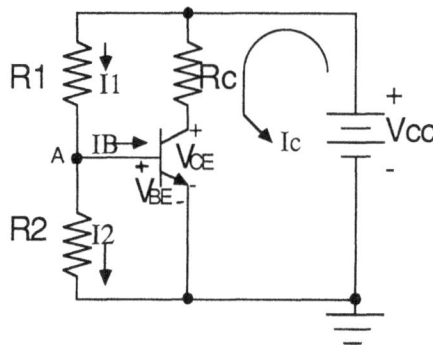

Fig. 4.10A

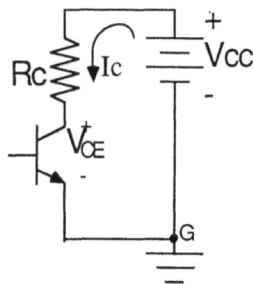

Fig. 4.10B

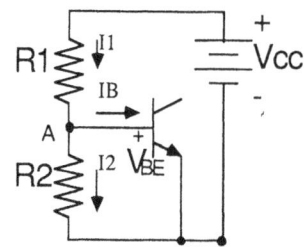

Fig. 4.10C

$R_2 = $ voltage across R_2/ current through R_2

and

$R_1 = $ voltage across R_1/ current through R_1.

The voltage across R_2 is defined by the transistor base to emitter voltage drop V_{BE}. The current through R_2 is simply I_2. We have

$R_2 = V_{BE}/I_2 = 1V/1.8mA = 556\Omega.$

The voltage across R_1 is the difference of two voltages. Note that the battery V_{CC} is on one side of R_1. The other side of R_1 is connected at point A which is going to the base. This voltage is V_{BE}. Consequently, the voltage across R_1 is the difference between V_{CC} and V_{BE}. The current through R_1 is simply I_1.

$R_1 = (V_{CC} - V_{BE})/ I_1 = (12V - 1V)/2mA = 5.56k\Omega.$

(4.9) For the following problems use the I/V chart in Figure (4.9) and the schematic in Figure (4.10A), calculate the unknown quantities in each problem. Note that all voltages are in volt, currents in mA, and resistances in ohm.

V_{CC}	R_C	V_{CE}	I_C	I_1	I_2	I_B	V_{BE}	R_1	R_2
20		5	40		$5I_B$		1		
18		6	30		$8I_B$		1		
25	500		40		$7I_B$		1.5		
15		8		$10I_B$		0.2	1.2		
20		10	10	$8I_B$			0.7		

(4.10) (Note: more difficult) For the following problems use the I/V chart in Figure (4.9) and the schematic in Figure (4.11). Calculate the unknown quantities in each problem. Note that the emitter current I_E is equal to the sum of the collector current I_C and the base current I_B, ie, $I_E = I_B + I_C$.

V_{CC}	R_C	V_{CE}	V_E	R_E	I_C	I_B	I_1	I_2	V_{BE}	R_1	R_2
25		10	5		20			$9I_B$	1		
20	200		3			0.3	$10I_B$		1.5		
18		7		150		0.4	$8I_B$		1		
20		8	4			0.2		$5I_B$	1.5		

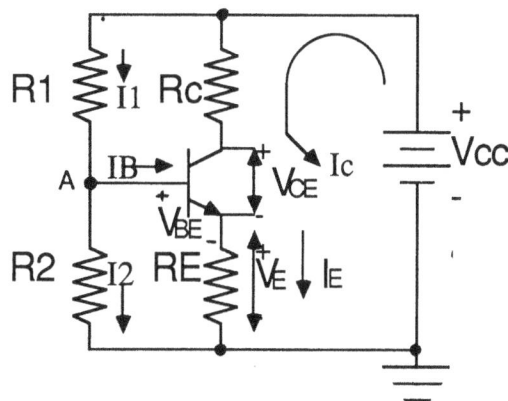

Fig. 4.11

4.1.5. Scattering Parameters

As discussed in Section 11.25 of the reference text, the two port scattering parameters for the network in Figure (4.12) are

S_{11} = input reflection coefficient = V_{1r} / V_{1i} (4.2A)

S_{21} = forward transmission coefficient = V_{1t} / V_{1i} (4.2B)

S_{22} = output reflection coefficient = V_{2r} / V_{2i} (4.2C)

S_{12} = reverse reflection coefficient = V_{2t} / V_{2i} (4.2D)

All of the following S-parameters are absolute magnitude.

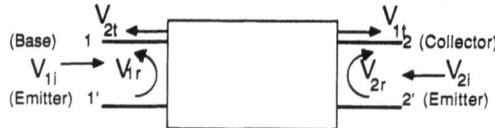

Fig. 4.12

Input return loss (dB) = $20 \log(1 / S_{11})$ (4.3A)

Input VSWR = $(1 + S_{11}) / (1 - S_{11})$ (4.3B)

Output return loss (dB) = $20 \log(1 / S_{22})$ (4.3C)

Output VSWR = $(1 + S_{22}) / (1 - S_{22})$ (4.3D)

Forward gain(dB) when $S_{21}>1 = 20 \log(S_{21})$ (4.3E)

Reverse gain (dB) when $S_{12}<1 = 20 \log(1/S_{12})$ (4.3F)

Isolation (dB) when $S_{12}<1 = 20\log(1/S_{12})$ (4.3G)

(4.11) For the following S-parameters, calculate input return loss, input VSWR, forward gain or loss, output return loss, output VSWR, and reverse gain or isolation. Also, define a common component which exhibit these properties

S_{11}	S_{21}	S_{22}	S_{12}	input RL(dB)	input VSWR	Forward gain(loss)dB	output RL(dB)	output VSWR	Reverse Gain(isol.)dB
0.4	2.0	0.25	0.2						
0.1	0.1	0.1	0.1						
0.2	0.89	0.15	0.05						

Calculation

We shall do the first one as an illustration.

Input return loss (dB) = $20 \log(1/0.4) = 20 \log(2.5) = 8dB$

Input VSWR = $(1 + 0.4) / (1 - 0.4) = 1.4 / 0.6 = 2.33$

Output return loss (dB) = $20 \log(1/0.25) = 20 \log(4) = 12dB$

Output VSWR = $(1 + 0.25) / (1 - 0.5) = 1.25 / 0.75 = 1.67$

Forward gain (dB) = $20 \log(20) = 26dB$

Isolation (dB) = $20 \log(1/0.2) = 14dB$

A device with forward gain and isolation is an ampifier.

4.1.7. Matching Circuits Using Inductors and Capacitors

This is a more difficult section and requires the understanding of the ZY chart which is shown in Figure (4.13). The readers are referred to the reference text for explanation of the ZY chart.

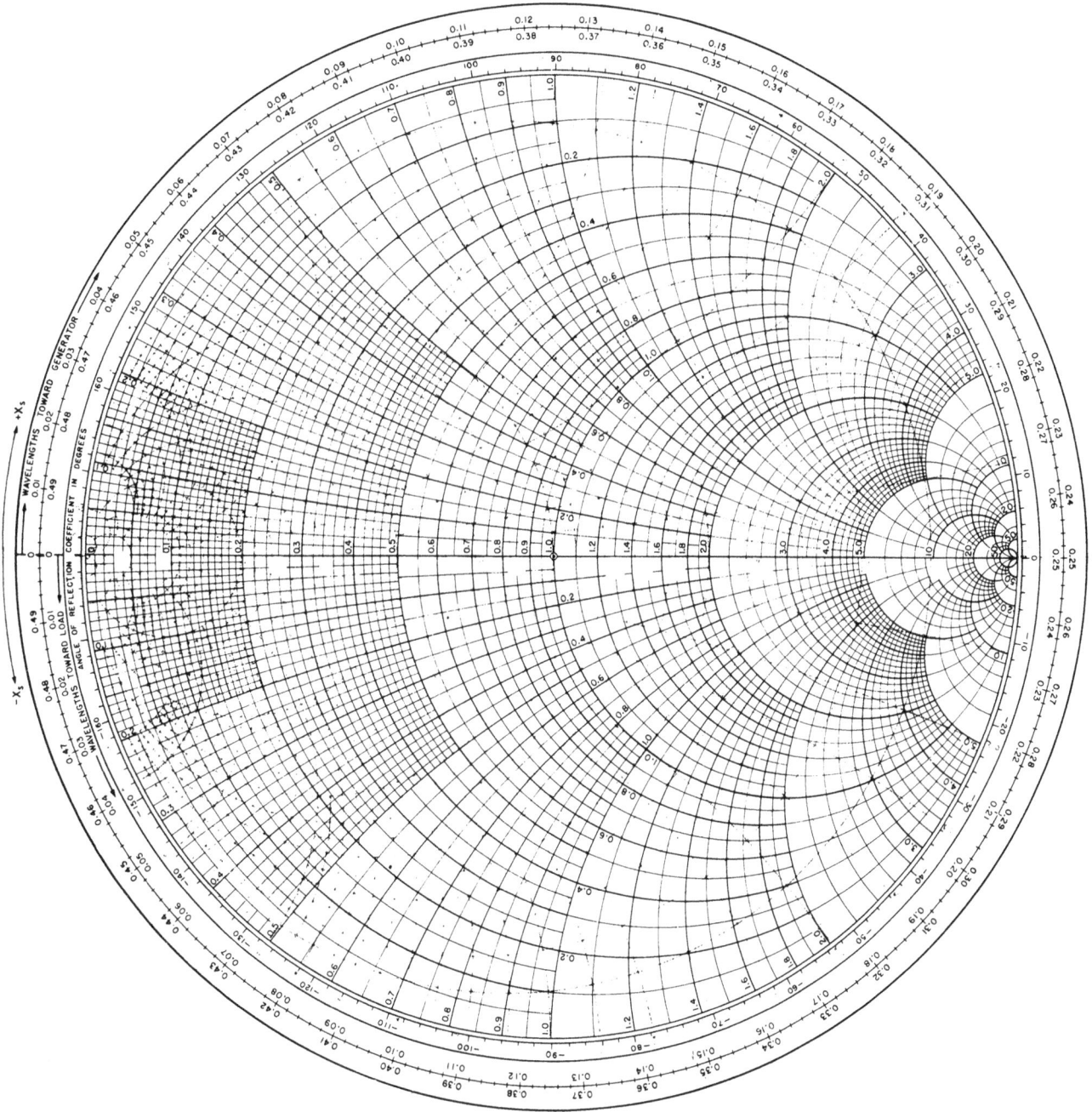

Fig. 4.13

(4.12)   The S_{11} of a transistor amplifier is 0.5-j1.5. for the input matching circuit of Figure (4.14A), calculate L_1 and L_2 at 1GHz for the shortest paths on the ZY chart such that $z_{in}=1.0+j0$. The characteristic impedance is 50Ω.

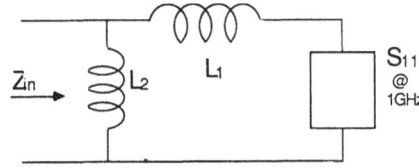

Fig. 4.14A

Calculation

Plot S_{11} using the 0.5 resistance circle and the 1.5 reactance curve in the lower half of the ZY chart in Figure (4.14B).

The L_1 path follows the 0.5 resistance circle and travels a z_1 distance,

$z_1 = 1.5 - 0.5 = 1.0$
$Z_1(actual) = z_1$ x 50Ω = 50Ω.

For a series inductor $Z_1 = 2\pi f L_1$, i.e.,

$L_1 = Z_1/2\pi f = 50Ω/2\pi \times 1\times10^9 Hz = 7.96nH$.

The shunt inductor L_2 follows the 1.0 conductance circle and travels a distance y_2,

$y_2 = 1.0 - 0 = 1.0$
Y_2 (actual) = $y_2/50Ω = 0.02mho$.

For a shunt inductor, $y_2 = 1/2\pi f L_2$, i.e.,

$L_2 = 1 /2\pi f Y_2 = 1 /2\times1\times10^9 Hz \times 0.02mho = 7.96nH$.

(4.13) The S_{11} of a transistor amplifier is 0.2-j1.4. For the input matching circuit of Figure (4.15A), calculate L_3 and L_4 at 1GHz for the shortest paths on the chart such that $z_{in} = 1.0+j0$.

Calculation

The S_{11} is plotted on a ZY chart in Figure (4.15B). The shunt inductor L_3 follows the 0.1 conductance circle and travels a y_3 distance,

$y_3 = 0.7 - 0.3 = 0.4$
$Y_3(actual) = y_3/50Ω = 0.4/50 = 0.008mho$
$L_3 = 1/2\pi f Y_3 = 1/2\pi \times 1\times10^9 \times 0.008 = 19.9nH$

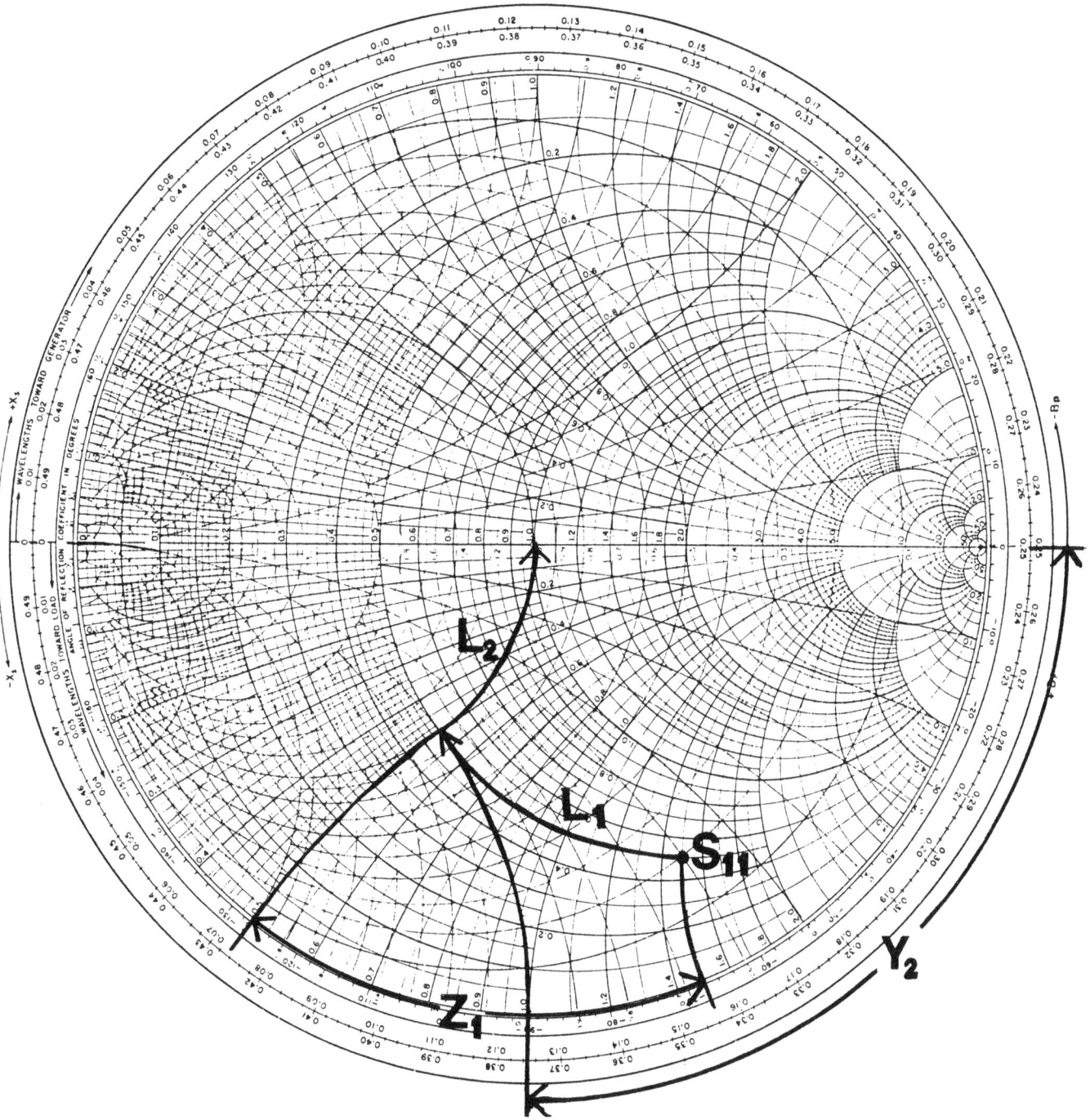

Fig. 4.14B

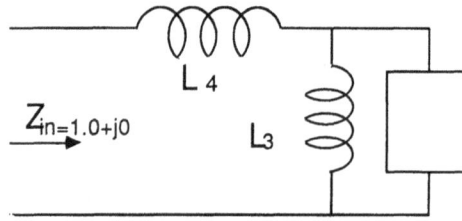

Fig. 4.15A

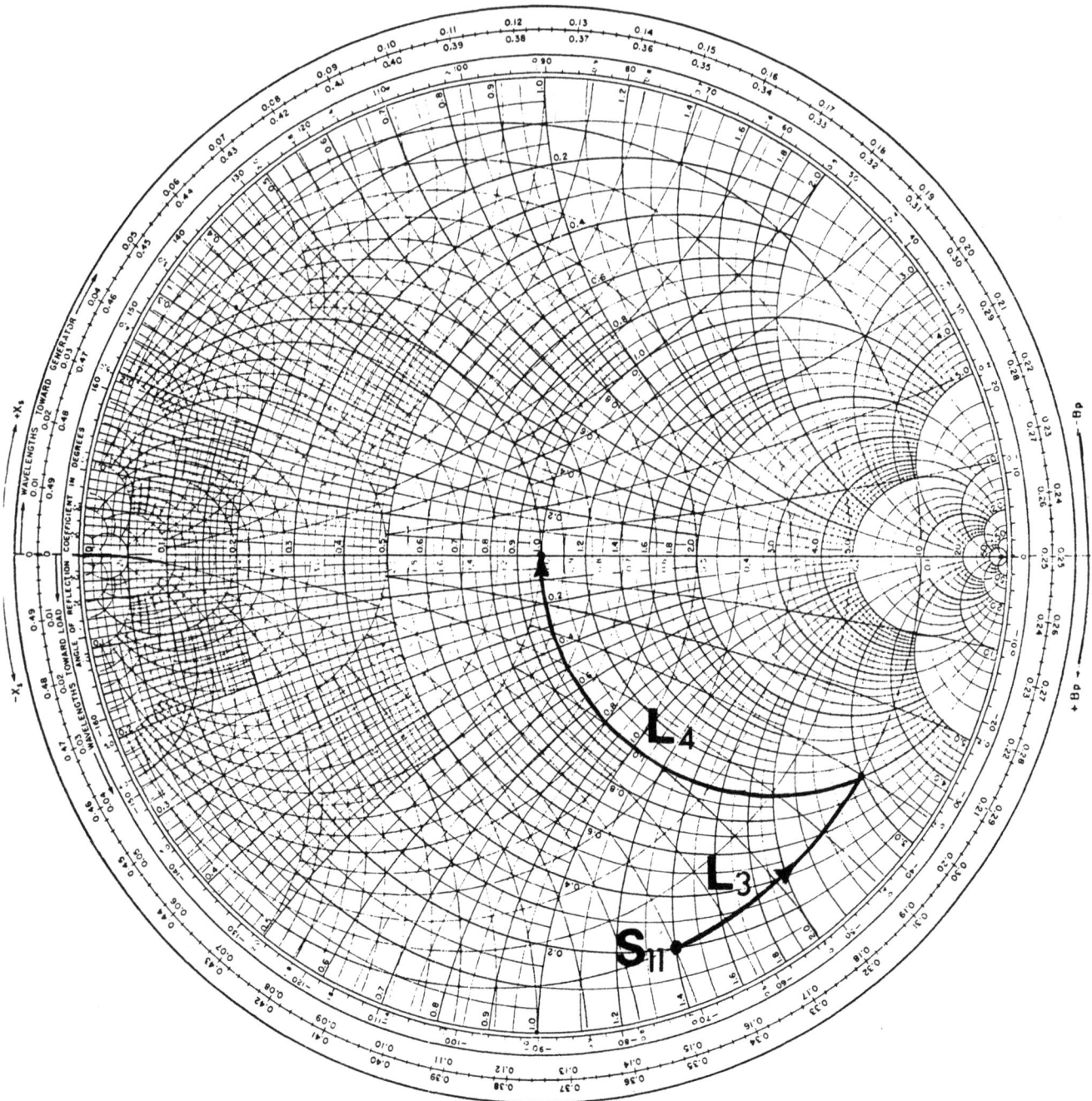

Fig. 4.15B

The series inductor follows the 1.0 resistance circle and travels a z_4 distance,

$z_4 = 3.0 - 0 = 3$
$Z_4(\text{actual}) = 3 \times 50\Omega = 150\Omega$
$L_4 = Z_4/2\pi f = 150/2\pi \times 10^9 = 23.9\text{nH}.$

(4.14) The S_{22} of a transistor amplifier is 2.5 -j2.5. for the output matching circuit of Figure (4.16A), calculate the value of C_1 and L_2 at 2GHz for the shortest paths on the ZY chart such that $z_{out}= 1.0+j0$

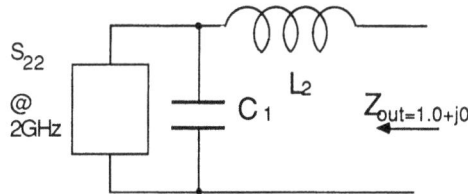

Fig. 4.16A

Calculation

The S_{22} is plotted on a ZY chart as shown in Figure (4.16B). The C_1 path follows the 0.2 conductance circle and travels a y_1 distance,

$y_1 = 0.4 - 0.2 = 0.2$
$Y_1(\text{actual}) = y_1/50\Omega = 0.2/50 = 0.004 \text{ mho}.$

For a capacitor, $Y_1 = 2\pi f C_1$, i.e.,

$C_1 = Y_1/2\pi f = 0.004 \text{ mho}/2\pi \times 2 \times 10^9 = 0.318 \text{ pF}.$

The series inductor L_2 follows the 1.0 resistance circle and travels a distance of z_2,

$z_2 = 2.0 - 0 = 2$
$Z_2(\text{actual}) = z_2 \times 50\Omega = 2 \times 50\Omega = 100\Omega$
$L_2 = Z_2/2\pi f = 100/2\pi \times 2 \times 10^9 = 7.96\text{nH}.$

(4.15) Use the same S_{22} in the previous problem and calculate the values of L_3 and C_4 at 2.0 GHz as shown in Figure (4.17) for the shortest paths on the ZY chart such that $z_{out} = 1.0 +j0$.

With reference to the ZY chart in Figure (4.16B) for the previous problem, the shunt inductor L_3 follows the 0.2 conductance circle and travels a y_3 distance,

$y_3 = 0.2 - (-0.4) = 0.6$
$Y_3(\text{actual}) = 0.6/50\Omega = 0.012 \text{ mho}$
$L_3 = 1/2\pi f Y_3 = 1/2\pi \times 2 \times 10^9 \times 0.012 = 6.63 \text{ nH}.$

The series capacitor C_4 follows the 1.0 resistance circle and travels a z_4 distance,

$z_4 = 2.0 - 0 = 2.0$

$Z_4 = 2.0 \times 50\Omega = 100\Omega$

$C_4 = 1/2\pi\ fZ_4 = 1/2\pi \times 2 \times 10^9 \times 100 = 0.796\ pF.$

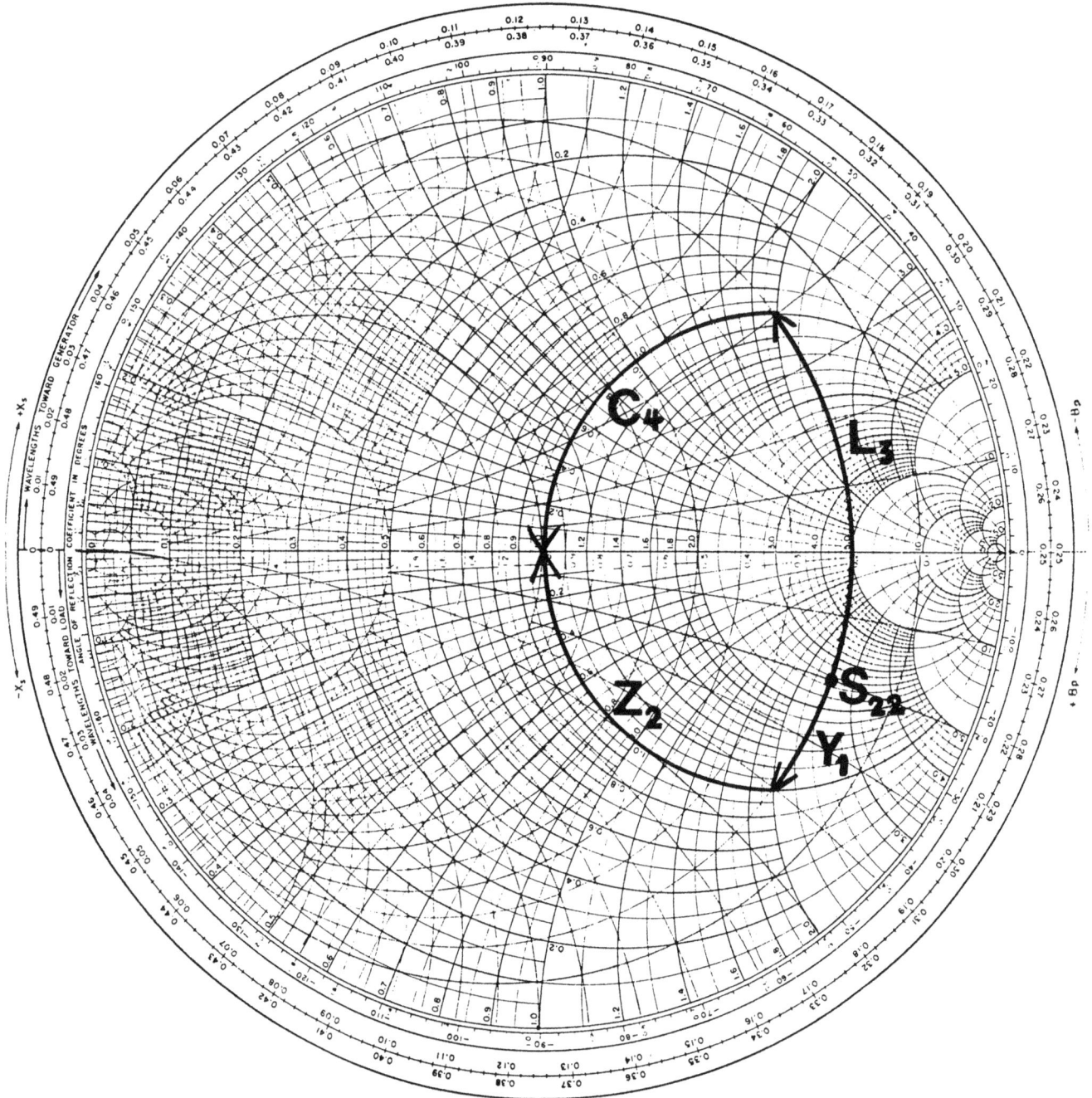

Fig. 4.16B

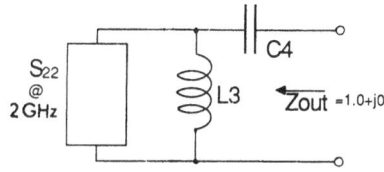

Fig. 4.17

(4.16)   Calculate the values of the matching elements using the shortest paths for the given normalized mismatches, circuits, and frequencies as shown in Figure (4.18).

a, $S_{11} = 0.5 - j1.0$, parallel and series inductor, f= 4GHz

b, $S_{11} = 0.5 - j1.0$, series and parallel inductor, f = 8GHz

c, $S_{22} = 0.2 + j0.2$, series inductor and parallel capacitor, f = 8GHz

d, $S_{22} = 0.2 + j0.2$, series capacitor and parallel inductor, f = 12GHz.

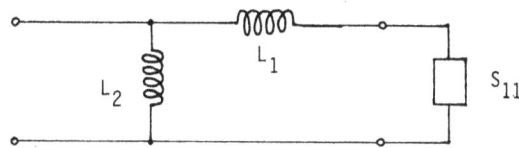

(A)

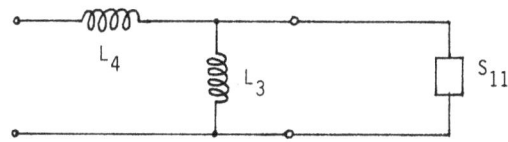

(B)

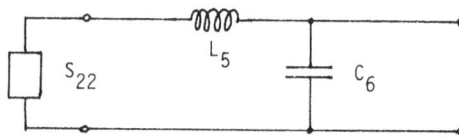

(C)

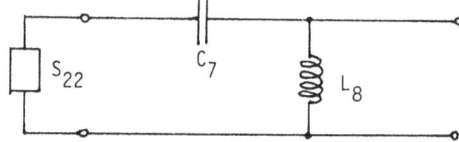

(D)

Fig. 4.18

4.2. FIELD EFFECT TRANSISTORS

4.2.1. Negative Gate Biasing

This biasing scheme requires the use of two power supplies as shown in Figure (4.19A). See Section 11.3.2 of the reference text.

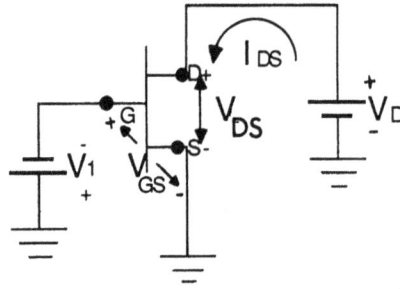

Fig. 4.19A

Exercise

(4.17)   For the circuit shown in Figure (4.19A), calculate V_{GS}, I_{DS}, and V_{DS} if $V_D = 5V$ and $V_1 = 1.5V$. Use the I/V chart in Figure (4.20).

Calculation

Step 1- Isolate the gate to source circuit first as shown in Figure (4.18B). The voltage drop V_{GS} is defined as the voltage at the gate V_G minus the voltage at the source V_S.

$$V_{GS} = V_G - V_S$$

Since $V_G = -V_1 = -1.5V$ and $V_S = 0V$ (the source is at ground), we have

$$V_{GS} = -1.5V - 0V = -1.5V$$

Step 2- Use the I/V chart in Figure (4.19) to find I_{DS} that gives $V_{GS} = -1.5V$,

$$I_{DS} = 25mA$$

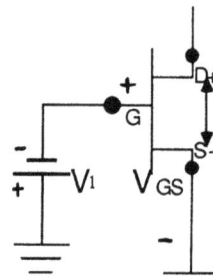

Fig. 4.19B

Step 3- Isolate the drain to source circuit as shown in Figure (4.19C).

The voltage drop V_{DS} is defined as the voltage at the drain V_D minus the voltage at the source V_S,

$$V_{DS} = V_D - V_S$$

Since $V_D = 5V$ and $V_S = 0V$ from above, we have

$$V_{DS} = 5V - 0V = 5V$$

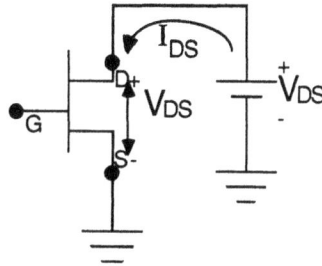

Fig. 4.19C

(4.18) For the exercise below, use the schematic in Figure (4.19A) and the I/V chart in Figure (4.20) to calculate the unknown quantities in each problem.

Solution:

$V_{GS}(V)$	$I_{DS}(mA)$	$V_1(V)$	$V_D(V)$	$V_{DS}(V)$
		1	10	
	15		8	
-3			5	
		2.5		6
	20		4	

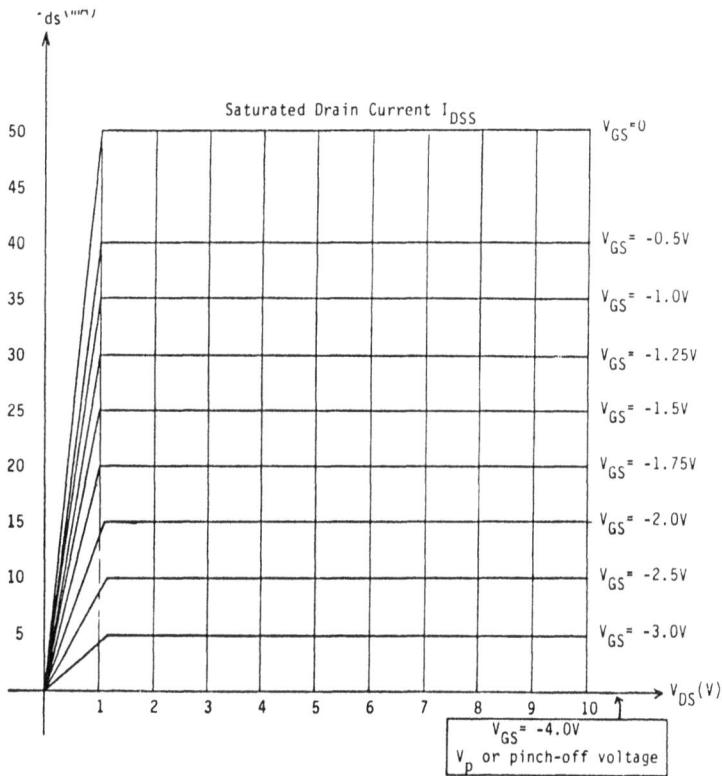

Fig. 4.20

4.2.2. Self Bias Mode

This biasing scheme uses only one power supply and one resistor as shown in Figure (4.21A)

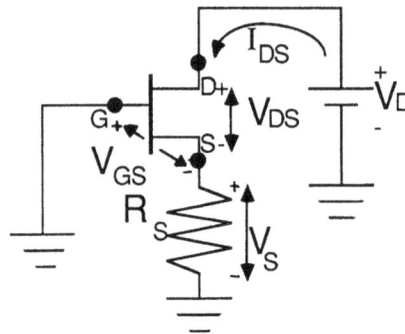

Fig. 4.21A

#Exercise#

(4.19) For the circuit in Figure (4.21A), $V_D = 5V$ and $I_{DS} = 10mA$ using the I/V chart in Figure (4.20), calculate V_{GS}, R_S, and V_{DS}.

Calculation

Step 1- Using the I/V chart in Figure (4.20), if $I_{DS} = 10mA$, then $V_{GS} = -2.5V$.

Step 2- Isolate the gate to source circuit as shown in Figure (4.21B).

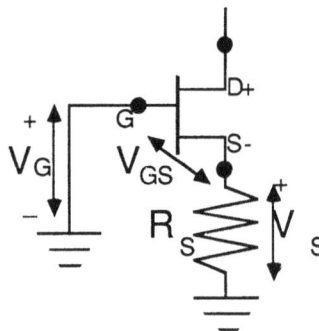

Fig. 4.21B

The voltage drop V_{GS} is defined as the voltage at the gate V_G minus the voltage at the source V_S,

$$V_{GS} = V_G - V_S$$

Since $V_G = 0V$ (ground) and $V_S = I_{DS}R_S = (10mA) R_S$, we have

$$-2.5 = 0 - (10mA)R_S$$
$$R_S = 2.5V/10mA = 250\Omega$$

Step 3-Isolate the drain to source circuit as shown in Figure (4.21C).

The voltage V_{DS} is the voltage at the drain V_D minus the voltage at the source V_S.

$$V_{DS} = V_D - V_S = 5V - (10mA)(250\Omega) = 2.5V$$

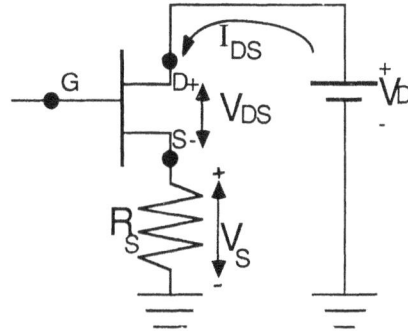

Fig. 4.21C

(4.20) For the problems below use the schematic in Figure (4.21A) and the I/V chart in Figure (4.20), calculate the unknown quantities in each problem.

$V_{GS}(V)$	$I_{DS}(mA)$	$R_S(\Omega)$	$V_D(V)$	$V_{DS}(V)$
	35		5	
-2			10	
	50% of I_{DSS}		12	
	10% of I_{DSS}		8	
			10	8.25
			7	5.75
			10	7.5
-0.5				4.5

Note 1: The saturated drain current I_{DSS} occurs when V_{GS} is zero volts on the I/V chart. Setting I_{DS} equal to one half the I_{DSS} means to make I_{DS} equal to one half the I_{DSS} or the current at which $V_{GS} = 0V$.

4.3 MICROSTRIP MATCHING CIRCUITS

See Section 11.3.9 of the reference text for detailed coverage of the microstrip circuits. A cross section of a microstrip transmission line is shown in Figure (4.22). The substrate of dielectric constant k is on a ground plane. The thickness of the substrate is h and a conductor of width w is on the top of the substrate. Microstrip transmission lines have a characteristic impedance Z_o determined by the ratio of thickness and width, i.e. Z_o is proportional to h /w.

#Exercise#

(4.21) For each of the microstrip matching circuit patterns in Figure (4.23), sketch the discrete equivalent circuit. Also, $Z_o = 50\Omega$ for h/w in each diagram.

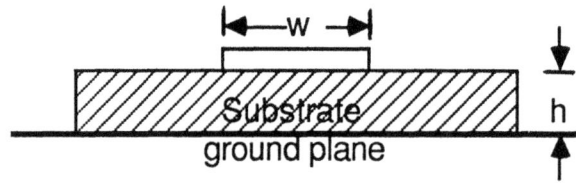

Fig. 4.22

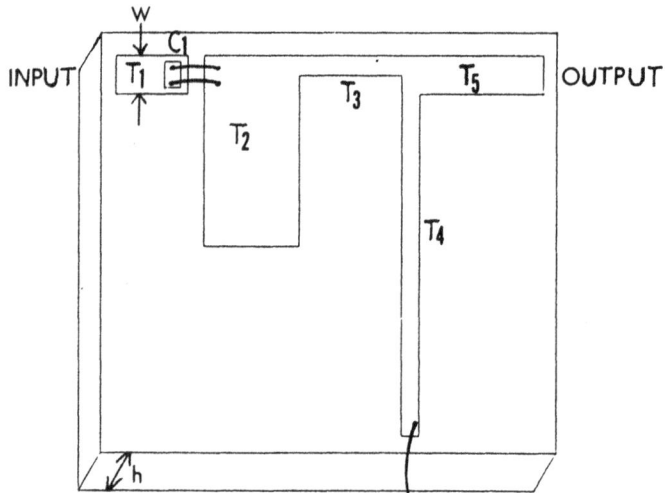

Fig. 4.23A

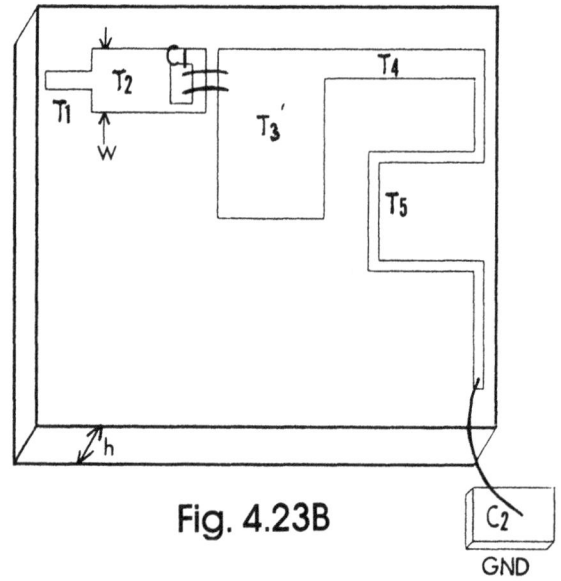

Fig. 4.23B

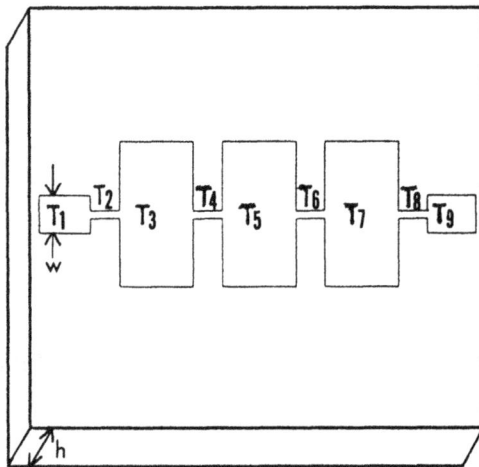

Fig. 4.23C

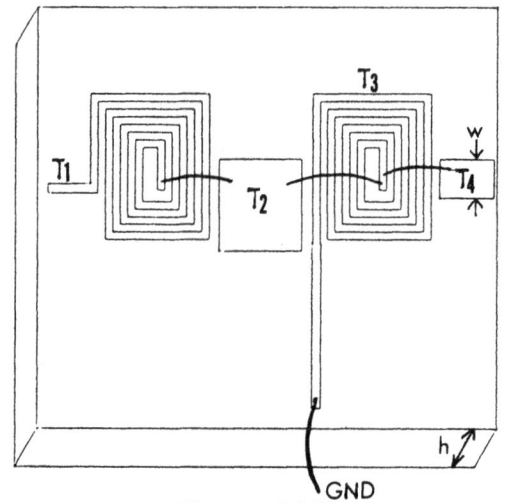

Fig. 4.23D

4.4. OSCILLATORS

There are several types of common resonators used in the microwave frequency range where the resonant frequency can be calculated by a formula. The formulas are derived from either an electrical condition of the circuit as with a lumped element resonant or the boundary conditions confining the fields as with a cavity resonator.

4.4.1 Common Resonators

A- Lumped Element Resonator (RLC), as shown in Figures (4.24A) and (4.24B).

$$f_{res} = 1/2\pi\sqrt{LC} \tag{4.4}$$

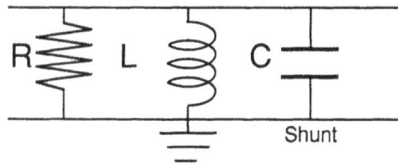

Fig. 4.24A

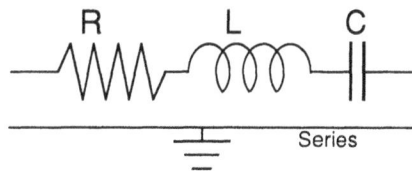

Fig. 4.24B

B- Cylindrical Cavity Resonator (air filled), as shown in Figure (4.24C).

$$f_{res} = \{c\sqrt{1 + [2L/3.41a]^2}\}/2L \tag{4.5}$$

where c is the speed of light, L is length and a is the radius as shown in the figure.

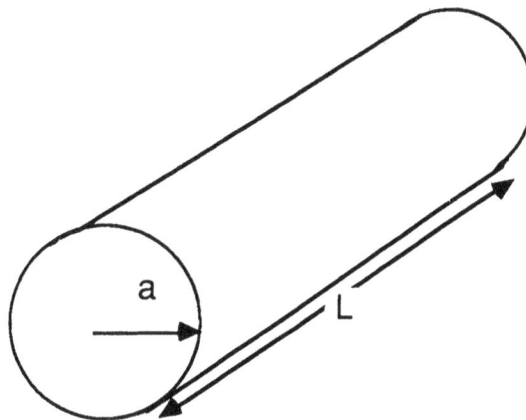

Fig. 4.24C

C- Rectangular Waveguide Resonator (air filled), as shown in Figure (4.25).

For b<a, $f_{res} = \{c \sqrt{a^2 + d^2}\} / 2ad$ (4.6)

where a and d are the width and the height of the waveguide.

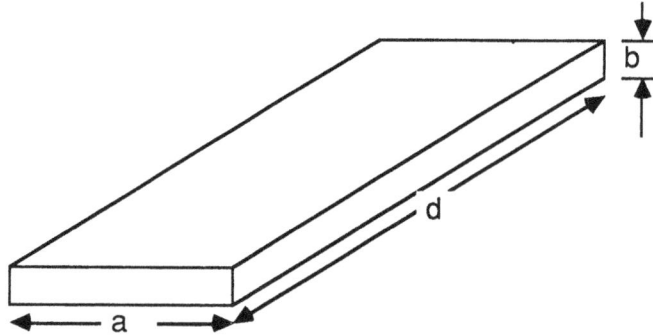

Fig. 4.25

#Exercise#

(4.22) Calculate the resonant frequency of a rectangular waveguide resonator where a= 1cm, d= 4cm, and b=0.5 cm. The speed of light c is 3×10^{10} cm/sec.

Calculation

It can be easily checked that b (0.5cm) is less than a (1cm). Use Equation (4.6) to find the resonant frequency,

$$f_{res} = 3 \times 10^{10} \sqrt{1^2 + 4^2} / 2(1)(4) \text{Hz} = 1.55 \times 10^{10} \text{ Hz} = 15.5 \text{ GHz.}$$

(4.23) Calculate the resonant frequency of an air filled cylindrical cavity resonator for the following dimensions:

a	L	f_{res} (GHz)
1 cm	4 cm	
2 cm	4 cm	
0.5 inch	1 inch	
1 inch	2 inch	
0.5 cm	3 cm	

(4.24) Calculate the resonant frequency of an air filled rectangular waveguide resonator for the following dimensions:

a	b	d	f_{res} (GHz)
2 cm	1 cm	10 cm	
4 cm	2 cm	10 cm	
2 inch	1 inch	4 inch	
1 inch	0.5 inch	2 inch	
1 cm	0.5 cm	2 cm	

(4.25) Calculate the resonant frequency of a lumped element resonator for the following parameters:

L (nH)	C (pF)	f_{res} (GHz)
1	1	
2	2	
5	5	
10	1	
0.5	0.5	

4.4.2. Varactor Tuned Oscillators (VTO) or Voltage Controlled Oscillators (VCO)

See Section 11.4.4 of the reference text for discussions in varactor tuned oscillators. Varactor tuned oscillators, sometimes called voltage controlled oscillators (VCO), are part of the negative resistance class of oscillators. The resonator is a series lumped element circuit (RLC). A varactor diode is a voltage variable capacitor as shown in Fig.4.26. This capacitor is part of the resonator along with an inductor and capacitor from a transistor.

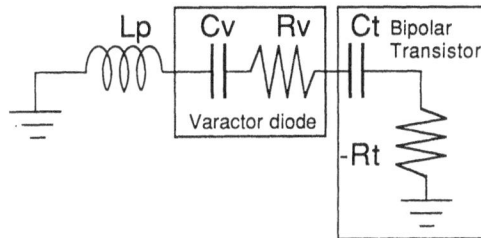

Fig. 4.26

The resonant frequency for this lumped element circuit is

$$f_{res} = 1/2\pi\sqrt{L_p C_{eq}} \qquad (4.7A)$$

where C_{eq} is the series combination of C_V and C_t, i.e.,

$$C_{eq} = C_V C_t /(C_V + C_t). \qquad (4.7B)$$

#Exercise#

(4.26) For the following varactor tuned oscillators, find the varactor capacitance and calculate the resonant frequency. Use the C_V versus V_R chart in Figure (4.27).

L_P (nH)	C_t(pF)	V_R(V)	C_V (pF)	f_{res} (GHz)
1	1	30		
1	1	15		
1	1	5		
0.5	0.5	50		
0.5	0.5	20		
0.5	0.5	10		

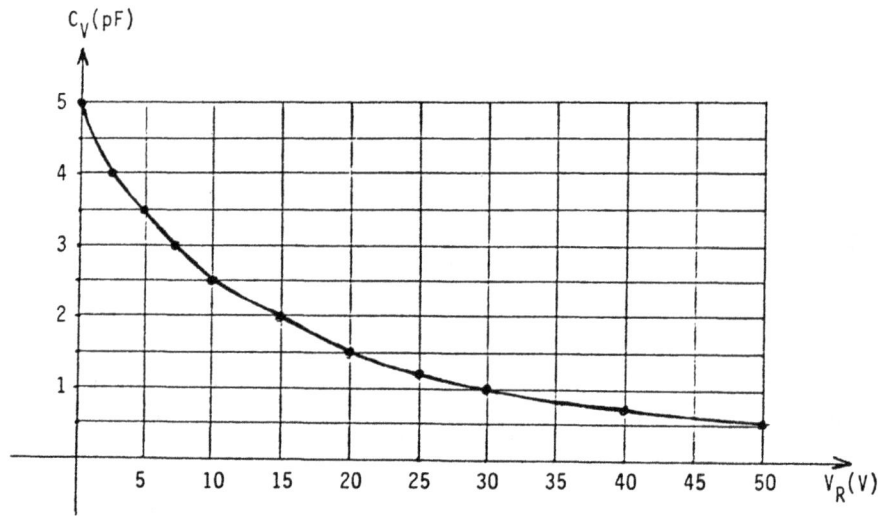

Fig. 4.27

5

MICROWAVE INTEGRATED CIRCUITS (MICs)

5.1 INTRODUCTION

This chapter seeks to reinforce microwave theory through the analysis of microwave integrated circuits in a hybrid form (MIC). All of the circuits used are derived from actual circuits in production today. Through the courtesy of the Devices Group of Watkins Johnson Company, located in Palo Alto, California, permission has been granted to publish these circuits. All of the circuits represent current technology and techniques used in high volume production.

This chapter requires a thorough understanding of Chapter Four on solid state amplifiers and oscillators. The readers is asked to sketch a schematic for each of the following exercises to aid an understanding of the MIC. By analyzing the schematic and the circuit diagram, insight can be gained into the DC biasing and RF tuning of the circuit.

5.2 BIPOLAR AMPLIFIERS

Figures (5.1) and (5.2) show circuits for bipolar transistor amplifiers. These amplifiers are built utilizing a substrate mounted onto a circular header called TO-8. These types of circuits operate in the frequency range of 10MHz to 2GHz with various bandwidths. Both of these circuits use negative feedback from the collector to the base to achieve a flat gain response over wide bandwidths. See Section 11.39 of the reference text for discussion of this topic.

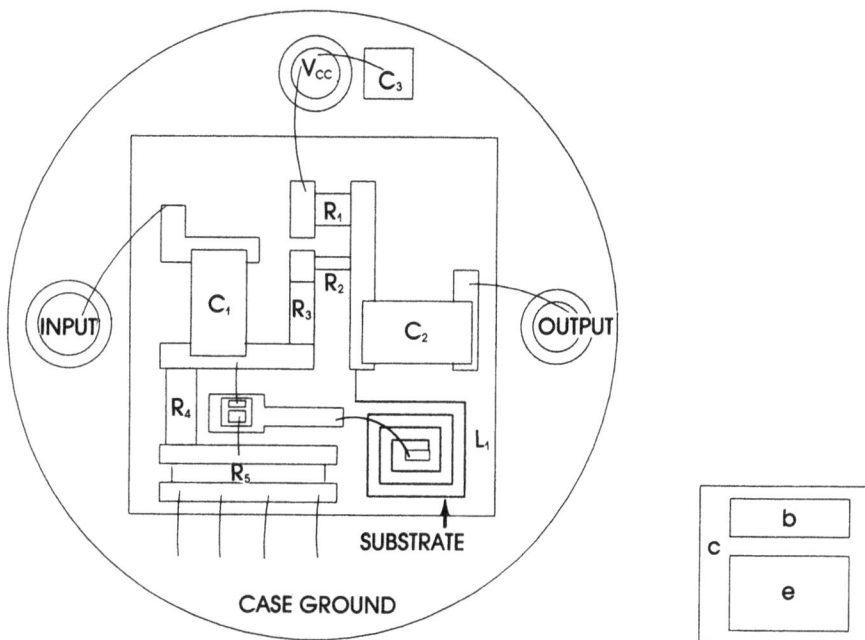

Fig. 5.1

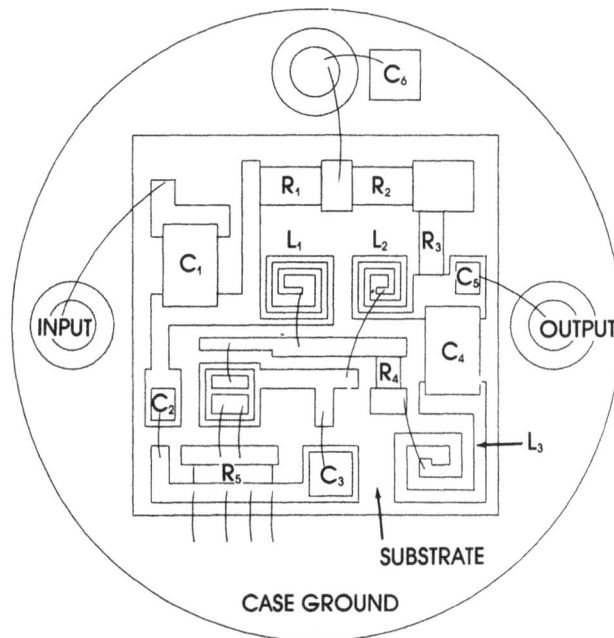

Fig. 5.2

Exercises

(5.1) In Figure (5.1), C_1 and C_2 are large coupling capacitors and C_3 is a bypass capacitor. Answer the following questions.

    a. Sketch a schematic representation of the circuit using resistors, capacitors, and inductors.
    b. Identify the elements which compose the dc bias network for the transistor.
    c. Identify the elements which compose the feedback network.
    d. Identify the elements of the output matching network.
    e. If R_3 were shorted, what would happen to the negative feedback and the gain of the amplifier. Assume the shift in the DC bias would have no effect on gain.
    f. (Note: more difficult) What is the purpose of R_5? What would happen if it were shorted?

(5.2) In Figure (5.2), C_1, C_4, and C_5 are large coupling capacitors. Also, C_6 is a large bypass capacitor. Both C_2 and C_3 are small in the range of 1 pF.

    a. Sketch a schematic representation of the circuit using resistors, capacitors, and inductors.
    b. Identify the elements which compose the dc bias network for the transistor.
    c. Identify the elements which compose the input matching network.
    d. Identify the elements which compose the output matching network.
    e. Identify the elements which compose the feedback network.
    f. (Note: more difficult) Explain what L_3 does to the feedback signal as a function of frequency.
    What effect does this have on the gain? What is the ultimate purpose of L_3? Remember that C_4 is simply a coupling capacitor and has no effect on the signal.

5.3 YIG TUNED OSCILLATORS

Figures (5.3) and (5.4) are YIG oscillator circuits. The static magnetic field produced by electromagnets (coils) on each side of the YIG sphere is not shown in the diagrams. The YIG sphere is mounted to a beryllium oxide, BeO, rod which is used to position the sphere in the static magnetic field. See Section 11.4.4 for discussion of this topic.

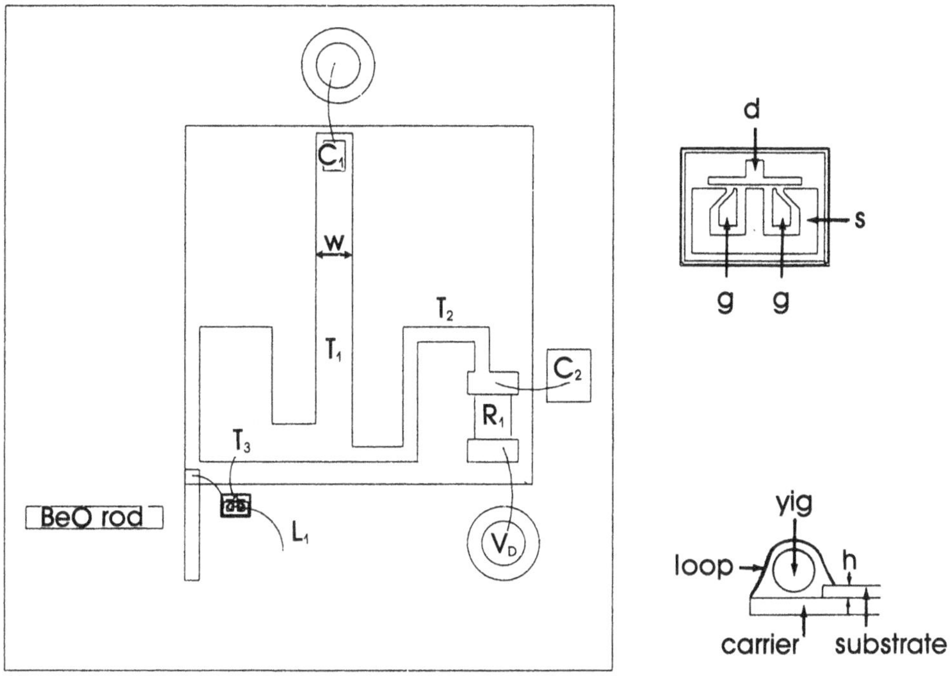

Fig. 5.3

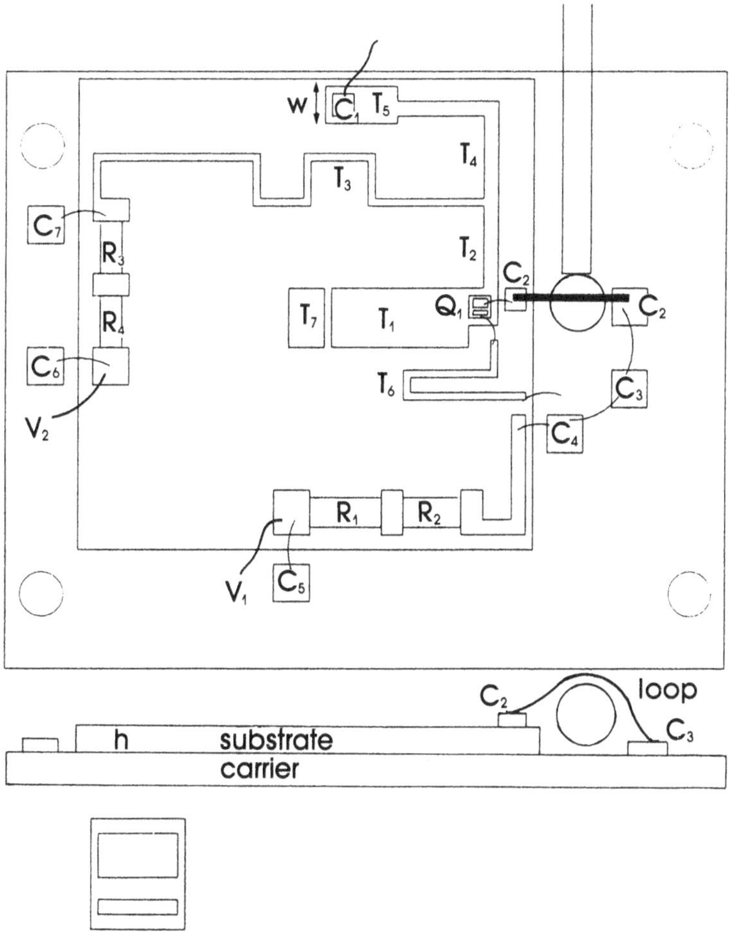

Fig. 5.4

Exercises

(5.3) Use Figure (5.3) to answer the following questions.

a. Sketch a schematic representation of the circuit. Assume that C_2 is a bypass capacitor and C_1 is a coupling capacitor.
b. Identify the elements which compose the YIG resonator.
c. Identify the elements which compose the output matching network.
d. What is the purpose of the FET and L_1 combination?
e. (Note: more difficult) The drain to source current, I_{DS}, is what percentage of the saturated drain to source current, I_{DSS}?

(5.4) Use Figure (5.4) to answer the following questions.

a. Sketch a schematic representation of the circuit. Capacitor C_1 is a coupling capacitor. Capacitors C_2 through C_7 are all bypass capacitors.
b. Identify the elements which compose the YIG resonator.
c. Identify the elements which compose the dc bias network for the transistor.
d. Identify the elements which compose the output matching network.
e. What is the purpose of the bipolar transistor and L_{T6} combination.
f. (more difficult) What would be the effect on T_1 if T_7 were connected to it?

5.4 VOLTAGE CONTROLLED OSCILLATORS

Figures (5.5) and (5.6) are varactor tuned, voltage controlled oscillator (VCO) circuits. These circuits are on a TO-8 metal header. The active device used is a bipolar transistor. See Section 11.4.4 for discussion of this topic.

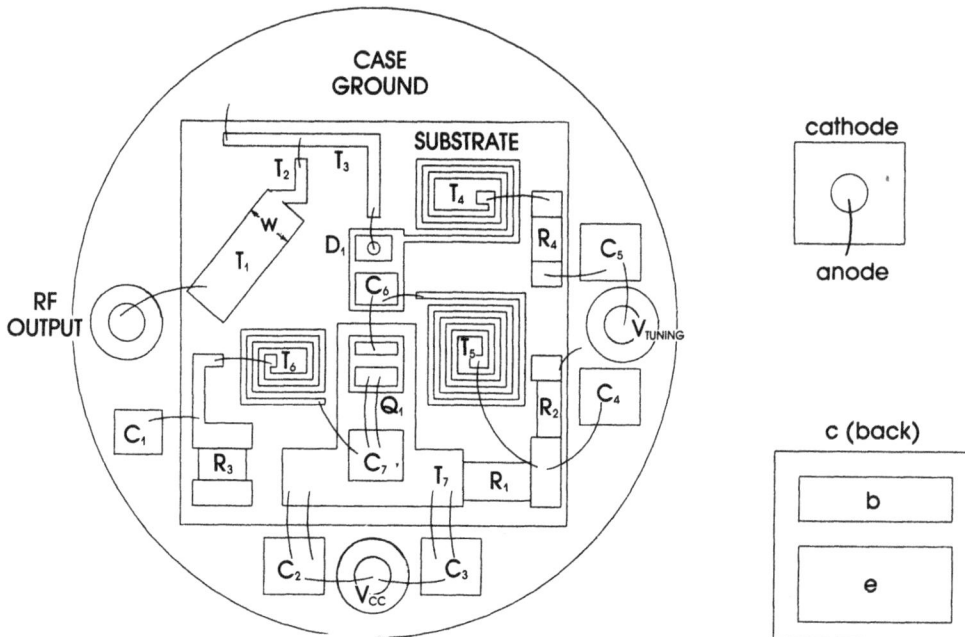

Fig. 5.5

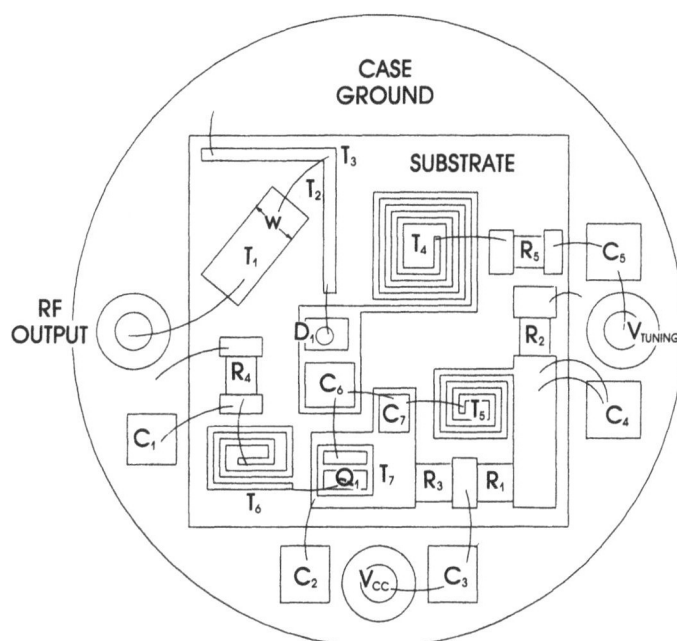

Fig. 5.6

Exercises

(5.5) Use Figure (5.5) to answer the following questions.

a. Sketch a schematic representation of the circuit. Capacitor C_1 through C_5 are bypass capacitors and capacitor C_6 is a coupling capacitor. Capacitor C_7 is a small value in the range 1pF. Hint: line T_2 "taps" line T_3.
b. Identify the elements which compose the dc bias network for the transistor.
c. Identify the elements which compose the resonator. What type of resonator is this?
d. What is the purpose of the transistor?
e. How is a signal coupled out of the resonator?
f. (Note: more difficult) How would you couple more energy out of the resonator?

(5.6) Use Figure (5.6) to answer the following questions.

a. Sketch a schematic representation of the circuit. Capacitors C_1 through C_5 are bypass capacitors and C_6 is a coupling capacitor. Capacitor C_7 is a small value in the range of 0.25pF.
b. Identify the elements which compose the dc bias network for the transistor.
c. Identify the elements which compose the resonator.
d. If V_T is a positive voltage, is the varactor diode forward or reverse biased?
e. As V_T increases, does the varactor capacitor, C_V, increase or decrease? Hint: Consult Figure (4.27) for this question.
f. What happens to the output frequency of the oscillator if the tuning voltage is made more positive?

5.5   FET AMPLIFIERS

Figures (5.7) and (5.8) are balanced FET amplifier circuits. These circuits operate in the 2 GHz to 20 GHz frequency range with bandwidths of 4-8 GHz, 8-12 GHz, 6-18 GHz, etc. The circuits are built on a metal carrier which is grounded. A rib usually made of golded plated copper is soldered to the carrier. The substrate material is usually alumina. Since the circuits in the upper and lower half are identical, it will only be necessary to sketch one half of the circuit. See Section 11.3.7 for discussion of this topic.

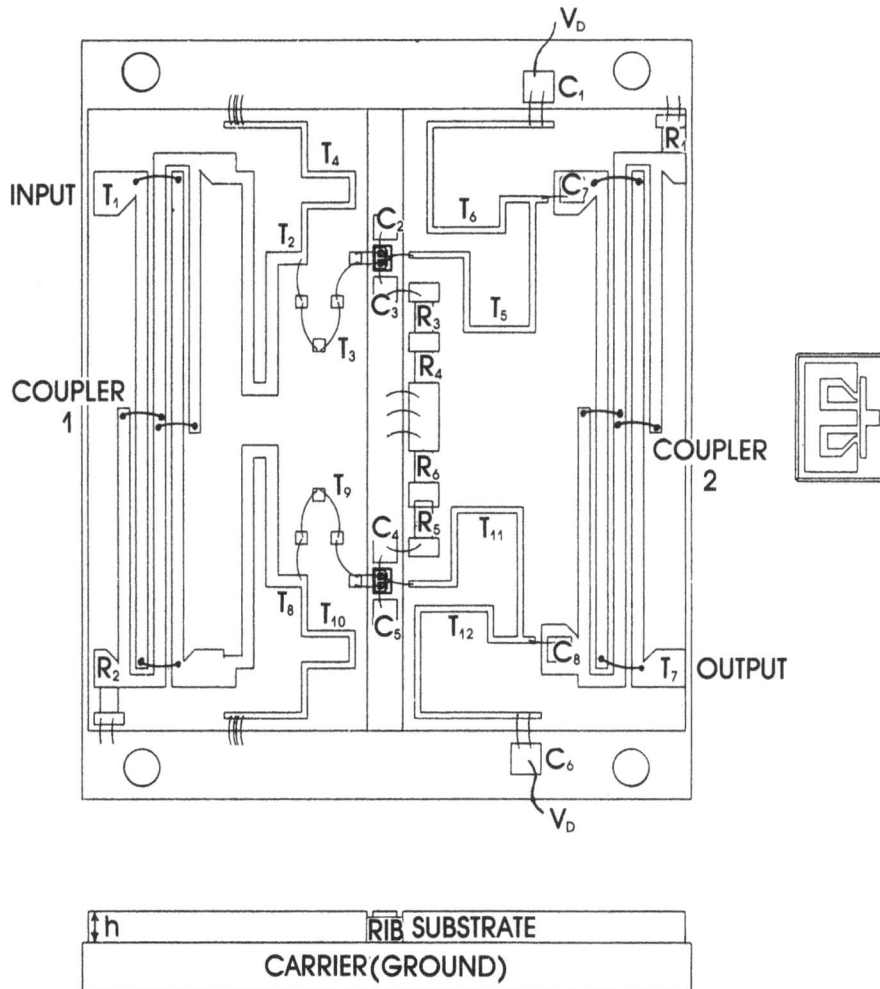

Fig. 5.7

Exercises

(5.7) Use Figure (5.7) to answer the following questions.

   a. Sketch a schematic representation of the circuit. Capacitors C_1 through C_6 are bypass capacitors. Capacitors C_7 and C_8 are small values. Resistors R_1 and R_2 are 50Ω.
   b. Identify the elements which compose the input matching network.
   c. Identify the elements which compose the output matching network.
   d. Is bias arrangement negative gate or self biased?
   e. What is the purpose of the couplers?

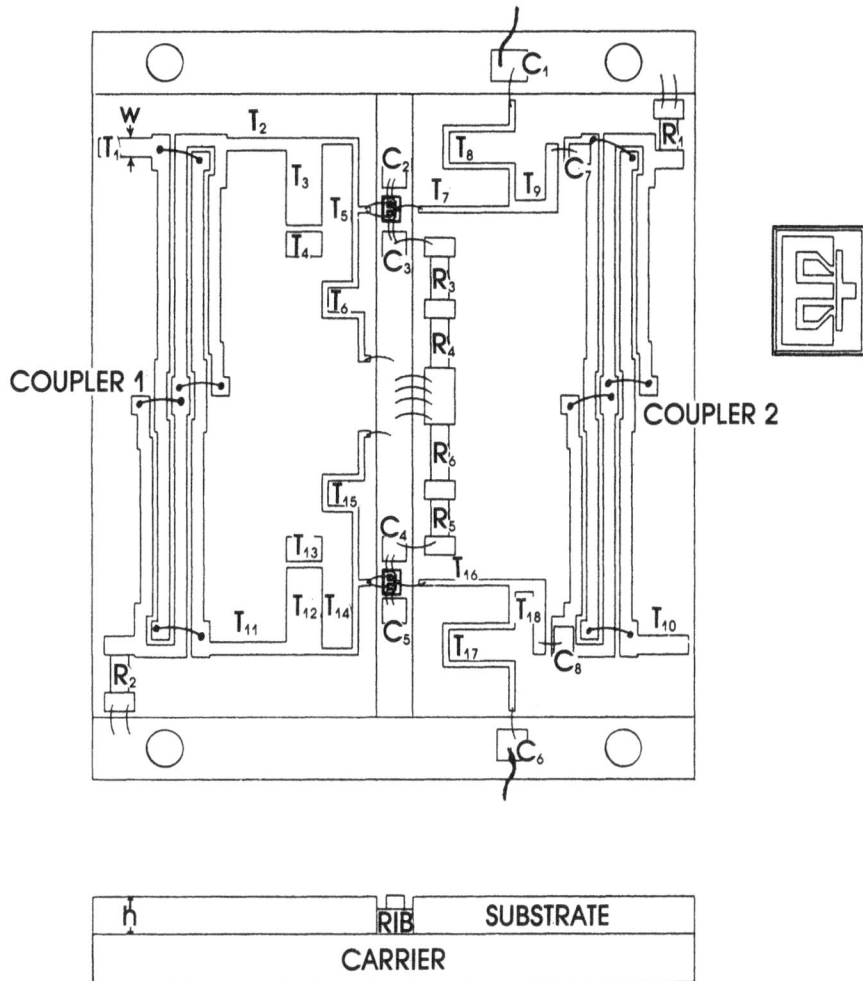

Fig. 5.8

f. (Note: more difficult) The DC bias can be adjusted by shorting R_3. Will the drain to source current, I_{DS}, increase or decrease?

g. (Note: more difficult) The RF matching can be tuned by modifying the bonding configuration. If the bonding configuration for T_3 is shorted, what will be the effect on L_{T3}?

(5.8) Use Figure (5.8) to answer the following questions.

a. Sketch a schematic representation of the circuit. Capacitors C_1 through C_6 are bypass capacitors. Capacitors C_7 and C_9 are coupling capacitors. Resistors R_1 and R_2 are 50Ω.

b. Identify the elements which compose the input matching circuit.

c. Identify the elements which compose the output matching circuit.

d. If T_4 were bonded to T_3 as part of RF tuning, what would be the effect on C_{T3}?

e. If you needed to increase I_{DS}, what would you do?

f. (Note: more difficult) Assume that at 8 GHz the match at T_2 and T_{11} is very poor such that one half the signal is being reflected. Where does this reflected power go? Hint: Consult Section 11.3.10.

g. (Note: more difficult) If R_3 and R_4 are shorted as an adjustment of DC bias, what would happpen to I_{DS}?

6

NOISE CALCULATIONS

6.1 DEFINITIONS

The following symbols are used in this chapter.

B = bandwidth (Hz)
G = circuit gain
$k = 1.38 \times 10^{-23}$ W/kHz (Boltzmann's constant)
NP = Noise power (watts)
NP_a = noise added by a circuit (watts)
NP_{in} = input noise power (watts)
NP_{out} = output noise power (watts)
NF = noise figure
T = temperature (Kelvins = Centigrade + 273)
T_e = equivalent noise temperature (Kelvins)
T_o = 290 K (standard noise temperature)
T_s = source temperature (Kelvins)

6.2 INTRODUCTION

Noise is discussed in a more extensive manner in Chapter 9 of the reference text.

The two fundamental limitations to the performance of microwave systems are random noise and signal distortion. The description of distortion is beyond the scope of this book but the effects of noise are easily understood with the help of a few mathematical principles.

Without random noise, there would be no lower limit to the power of detectable signals. No signal would be too weak to amplify and detect. We could communicate across the continent, or across the solar system, with transmission signal powers of only a few milliwatts. Portable communicators supplied by miniature batteries could communicate via satellite clearly and distinctly.

The reality is quite different. To communicate across long distances, people have to develop high power transmitters and elaborate antennas that are bulky and definitely not portable. In the following problems and examples, you will practice using a few simple equations that describe the effects of noise in microwave circuits.

6.3 NOISE TEMPERATURE

Every electrical component generates random noise due to the thermal vibration of its atoms and electrons. These vibrations generate minute voltages and currents, which are added to any signal present in the device. The total noise power NP available from an electrical component at a temperature T and over a bandwidth B is equal to

$$NP = kTB \qquad (6.1)$$

This shows that noise power depends on temperature and bandwidth. Because of the temperature dependence, a reference temperature has been established for noise calculations. This temperature is called T_O and has the value $T_O = 290$ K.

The noise power in a 1 Hz bandwidth due to T_O is equal to

$$NP_O = NP(290K, 1Hz) = 4 \times 10^{-21} \text{ W} = -174 \text{ dBm.}$$

In a 1 MHz bandwidth circuit, the total noise power from a source at T_O is

$$NP (290 \text{ K, 1 MHz}) = 4 \times 10^{-15} \text{ W} = 0.004 \text{ pW} = -114 \text{ dBm.}$$

In a circuit with a bandwidth of 1 GHz, the same equivalent noise power is 4.0 pW.

To find the total noise power at T_O in a certain bandwidth B, convert the bandwidth to equivalent dB by taking 10 log(bandwidth in Hz) plus NP_O.

$$NP(290K, B) = [-174 + 10 \log(B \text{ in Hz})] \text{ dBm} \qquad (6.2)$$

Exercises

In all the following problems, use the reference temperature of 290 K unless otherwise instructed.

(6.1) Calculate the noise power in dBm for a 40 MHz bandwidth at T_O.

Calculation

$$NP = -174 + 10 \log(40 \times 10^6) = -174 + 76 = -98 \text{ dBm}$$

(6.2) Use Equation (6.2) to calculate the equivalent noise power for the following bandwidths.

    a. 50 Hz (low speed, deep space telemetry)
    b. 4 kHz (telephone circuit)
    c. 50 kHz (low speed data link)
    d. 4 MHz (television)
    e. 36 MHz (satellite transponder)
    f. 500 MHz (ultra-high-speed PCM receiver)
    g. 10 GHz (wideband EW receiver)

Notice that the range of available noise power varies widely depending on the bandwidth of the application.

Calculation

    a. NP = -174 + 10 log (50) = -157 dBm

(6.3) The noise power available from thermal sources at 290 K in a 1 MHz bandwidth is -114 dBm. This value will be wrong for equipment not at exactly 290 K. What is the correct value for temperatures from 0 to 50 C?

Calculation

Since 0° C = 273K and 50^oC = 323K,

NP(273K, 1MHz) = $kx273x10^6$ = $3.76x10^{-15}$ W = 0.00376 pW = -114.2 dBm
NP(323K, 1MHz) = $kx323x10^6$ = $4.45x10^{-15}$ W = 0.00445 pW = -113.5 dBm

The difference in noise power is small.

A source of random noise power NP can be expressed as equivalent temperature T_e where

$$T_e = NP/kB \qquad (6.3A)$$

This is more easily expressed as ratio,

$$T_e = T_o (NP /NP_o B) \qquad (6.3B)$$

(6.4) If a circuit supplies 2.5 nW of noise in a 100 MHz bandwidth, find the equivalent noise temperature.

Calculation #:

Using Equation (6.3B),

$$T_e = 290 \times (2.5 \times 10^{-9}/4 \times 10^{-21} \times 100 \times 10^6) = 1,812,500 \text{ K!}$$

That is the (hypothetical) temperature that would produce 2.5nW noise power in a 100 MHz bandwidth.

(6.5) Calculate the equivalent noise temperature for the following exercises, given the values for noise power and bandwidth.

a. 220 pW, 2 GHz
b. 6.6 pW, 800 MHz
c. 0.05 pW, 50 kHz
d. 350 pW, 2 MHz
e. 30 pW, 20 GHz
f. 1 pW, 50 MHz

Sample calculation

a. T_e = 290 x 220 x 10^{-12}/4 x 10^{-21} x 2 x 10^9 = 7975 K

6.4 SIGNAL-TO-NOISE RATIO AND NOISE FIGURE

Signal-to-noise ratio is defined as

$$S/N = \text{Signal Power} / \text{Noise Power} \qquad (6.4A)$$

In decibel,

$$S/N(dB) = \text{Signal power (dBm)} - \text{Noise power (dBm)} \tag{6.4B}$$

(6.6)	Signal power	Noise power	S/N(#)	S/N(dB)
	5nW	1pW		
	1pW	0.005pW		
	20pW		500	
	-90dBm			30

The basic definition of the noise figure of a circuit is

$$NF = (S/N)_{in} / (S/N)_{out} \tag{6.5A}$$

provided that the equivalent noise temperature of the source is at T_o. The careless application of Equation (6.5A) can lead to mistakes. There are many situations where the equivalent noise temperature at the input to an amplifier is not equal to T_o.

Equation (6.5A) can also be expressed in dB as

$$NF(dB) = S/N(dB)_{in} - S/N(dB)_{out} \tag{6.5B}$$

(6.7)	Input S/N(dB)	Output S/N(dB)	NF(dB)
	30	20	
	53	45	
	35		10
		26	8

If the input noise is not at a 290 K equivalent, then the ratio of signal to noise ratios is

$$\frac{(S/N)_i}{(S/N)_o} = 1 + \frac{290}{T_s}(NF - 1) \tag{6.6}$$

where T_s is equivalent temperature at the source and can be calculated from the input noise power from Equation (6.3A) or (6.3B).

Note that Equation (6.6) reduces to Equation (6.5A) when $T_s = 290$ K.

In general, if the equivalent source temperature is greater than 290 K, the degredation in S/N ratio is less than the value of the noise figure. However, if $T_s < 290$, then the change in S/N ratio is greater than the value of NF.

(6.8) Calculate the output S/N ratio in dB for the following situations. Use Equation (6.6) and remember to convert dB terms to numerical.

	$(S/N)_{in}$	T_s (K)	NF(dB)
a.	25 dB	125	4.0
b.	30 dB	600	7.0
c.	6 dB	5,000	25.0

Sample calculation

a. First, we convert $(S/N)_i$ and NF to numerical values.

$(S/N)_i$ = 25 dB = 316 (an exact value is chosen here) and NF = 4 dB = 2.5. Using Equation (6.6) to solve for $(S/N)_o$, we have

$$\frac{316}{(S/N)_o} = 1 + \frac{290}{125}(2.5 - 1)$$

which results in

$$(S/N)_o = 70.1 = 18.5 \text{ dB}$$

(6.9) (Note: more difficult) A communications channel has a bandwidth of 40 MHz. The input power to the transmitter amplifier is +3 dBm at a S/N ratio of 46 dB. The amplifier has a noise figure of 40 dB. What is the output S/N ratio from the amplifier? Hint: Calcualte the input noise power from the S/N ratio and convert to T_s.

Calculation

From the definition of $(S/N)_i$ in decibel form,

$$NP_{in} = P_{in} - (S/N)_{in} = 3 \text{ dBm} - 46 \text{ dB} = -43 \text{ dBm} = 5 \times 10^{-5} \text{ mW}$$

Using Equation (6.3B), we have $T_S = 9 \times 10^7$ K.
We can now solve for $(S/N)_o$ using Equation (6.6), noting that NF = 40 dB = 10,000

$$\frac{4 \times 10^4}{(S/N)_o} = 1 + \frac{290}{9 \times 10^7}(10^4 - 1)$$

which we obtain $(S/N)_o = 3.875 \times 10^4 = 45.88 \text{dB}$.

(6.10) The circuit in Figure (6.1) represents an amplifier followed by a detector. For each combination of the following noise figures and bandwidths, calculate the input power required to obtain an output S/N of 3dB. Hint: Solve for the input S/N ratio in terms of the input power and the circuit bandwidth

	NF	Bandwidth
a.	5 dB	1 MHz
b.	6 dB	75 MHz
c.	2.5 dB	150 MHz
d.	15 dB	300 MHz
e.	35 dB	300 MHz

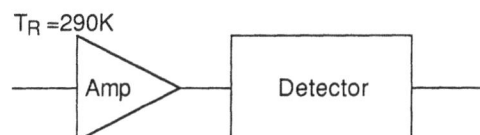

Fig. 6.1

#Sample solutions #

a. Noting that NF = 5 dB = 3.16 and the equivalent input noise temperature is 290 K, use Equation (6.6) to solve for the input S/N.

$$\frac{(S/N)_i}{(S/N)_o} = 1 + \frac{290}{290}(3.16 - 1) = 3.16$$

Since $(S/N)_o$ = 3 dB = 2, we can calculate $(S/N)_i$

$$(S/N)_i = 2 \times 3.16 = 6.32 \ (= 8.0 \ dB)$$

The equivalent thermal noise power is -174 dBm/Hz + 10 log(10^6) = -174 + 60 = -114 dBm = 4×10^{-15} W. Therefore, the input power is

$$P_{in} = (S/N)_i \times noise \ power = 6 \times 4 \times 10^{-15} \ W = 2.4 \times 10^{-14} \ W$$

6.5 ADDED NOISES

Noise figure describes how much noise a circuit adds to the signal. In Figure 6.2, the output noise NP_{out} is equal to the amplified input noise plus NP_a, the added noise.

$$NP_{out} = GNP_{in} + NP_a \tag{6.7}$$

Or since $NP_{in} = kT_SB$,

$$NP_{out} = kT_sGB + NP_a \tag{6.8}$$

Note that both equations are in numerical form, not decibel.

In terms of the noise figure of the circuit, the added noise is given without proof to be

$$NP_a = kT_oGB(NF - 1) \tag{6.10}$$

This is due to an alternative (but completely equivalent) definition of noise figure which states

$$NF = \frac{Total \ output \ noise \ power}{Noise \ power \ due \ to \ source} \times (source \ at \ 290 \ K) \tag{6.11}$$

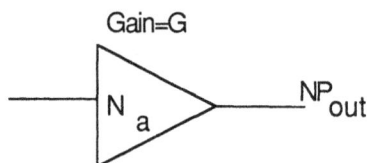

Fig. 6.2

(6.11) Calculate the added noise for the following combinations of gain, bandwidth, and noise figure.

	G(dB)	B	NF(dB)
a.	15	3MHz	5
b.	20	30kHz	1.5
c.	45	4MHz	0.5
d.	6	10GHz	8

Sample calculation

a. Convert dB to numerical (NF= 5dB = 3.16 and G =15 dB = 31.6), and use Equation (6.10).

$$NP_a = (1.38 \times 10^{-23}) \times (290) \times (3 \times 10^6) \times (31.6) \times (3.16 - 1) = 8.2 \times 10^{-13} \ W = 90.9 \ dBm$$

6.6 EQUIVALENT NOISE TEMPERATURE

An alternate description of a noisy circuit is its equivalent noise temperature, T_e, which is related to the noise figure as follows. Equivalent noise temperature has been discussed briefly in Section 6.3.

$$T_e = T_o (NF - 1) \tag{6.12}$$

or

$$NF = \frac{T_o + T_e}{T_o} \tag{6.13}$$

The significance of T_e is shown in Figure (6.3). If NP_{out} is the output noise of a noisy amplifier with a source temperature T_s, then T_e is the excess temperature at the source that would produce the same output noise power if the amplifier was noiseless.

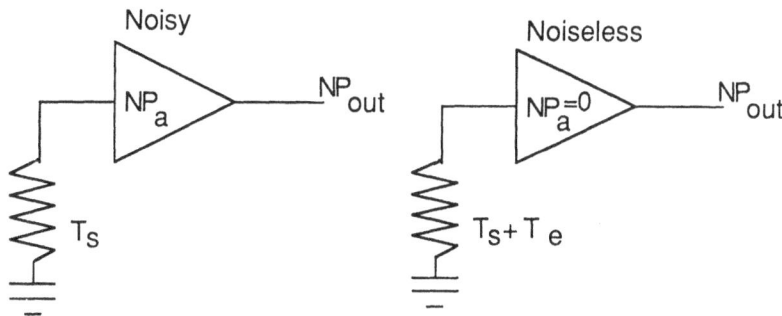

Fig. 6.3

(6.12) Convert the following values of nosie figure to equivalent noise temperature, and vice versa.

	NF(dB)	T_e
a.	10	-----
b.	1	-----
c.	-----	500
d.	-----	1,500
e.	17	-----
f.	3	-----
g.	-----	35
h.	-----	0

What is the significance of the result for (h)?

Sample calculation

a. NF = 10dB = 10
$T_e = 290 \times (10-1) = 2610K$

6.7 CASCADED NETWORKS

The noise figure of two or more circuits connected in cascade is,

$$NF = NF_1 + \frac{NF_2 - 1}{G_1} + \frac{NF_3 - 1}{G_1 G_2} + \ldots\ldots \quad (6.14)$$

where NF_1, NF_2, etc are the noise figures of the first, second, etc. circuits, and G_1, G_2, etc. are the gains. These are numerical values of NF and G, not dB values.

The most important result of Equation (6.14) is that the noise figure of the second and subsequent stages do not matter much if the gain of the first circuit is high.

Exercises

(6.13) Calculate the overall noise figure in dB of a series combination of amplifiers given the following parameters.

	NF_1(dB)	G_1(dB)	NF_2(dB)
a.	5.0	10.0	12.0
b.	12.0	30.0	2.5
c.	2.4	36.0	15.0
d.	6.0	0.0	7.5

Sample calculation

a. Convert the quantities in dB into numerical values. To be a little more precise, we use NF = 5.0dB = 3.16, G = 10dB = 10, NF = 12dB = 15.85. Use Equation (6.14) to find the total noise figure.

$$NF = 3.16 + (15.85 - 1)/10 = 4.645 = 6.67dB$$

7

ANTENNAS AND PROPAGATION

7.1 DEFINITIONS

S = distance from antenna
A = antenna aperture
f = frequency
G = gain
G_r = receiving antenna gain
G_t = transmitting antenna gain
HPBW = half power beamwidth
P_d = power density
P_t = transmitted power
P_r = received power
L = wavelength
D = size of antenna (usually the diameter)

7.2 INTRODUCTION

Antennas are covered in Chapter 13 of the reference text. Antennas are structures for transferring electromagnetic waves between free space and a transmission line. The transfer is in either direction; either receiving or transmitting. In fact, any antenna can act as either type of device.

The properties of antennas such as polarization, beamwidth, gain, side lobe suppression, and others, can be measured using various techniques that usually involve moving around a smaller "probe" antenna to measure the properties of the beam and its side lobes. Sometimes the "probe" antenna is held still while the main antenna is rotated about, placing the probe in various portions of the beam. The net effect is the same.

7.3 NEAR FIELD/FAR FIELD

The space around an antenna is divided into two regions; the near field and the far field. In the near field, the radiation from an antenna is distorted and has not acheived its final geometric form. In the far field, the beam is well behaved and predictable.

The approximate distance from an antenna to the far field region is

$$d = 2D^2/L \tag{7.1}$$

Exercises

(7.1) Calculate the distance to the far field for the following antennas.

a. Parabolic, dia. = 3 m, f = 4 GHz
b. Parabolic, dia. = 10 m, f = 12 GHz
c. Parabolic, dia. = 30 m, f = 1.2 GHz

Sample calculation

a. Frequency = 4GHz means that the wavelength is 7.5cm = 0.075m

Far field distance d = $2 \times 3^2 / 0.075 = 240$m

(7.2) Possibly the largest parabolic antenna in the world is the parabolic reflector at Arecibo, Puerto Rico. Operated by Cornell University, it is used for radio astronomy research. It is a fine metallic mesh suspended between surrounding hilltops in a natually bowl-shaped valley. The diameter of the dish is approximately 300 meters. Calculate the distance to the far field at operating frequencies of a) f = 500 MHz, b) f = 2 GHz

(7.3) Measuring the shape of the antenna beam for the Arecibo observatory could pose quite a problem. Suppose you fly over the beam in an airplane at 30,000 feet with a small receiving antenna. What frequency should you use to be sure of getting accurate results? Hint: remember the plane should in the far field region.

Calculation

Set the far field d = 30,000ft = 9000m. We then solve for the wavelength knowing that the diameter of the antenna is D = 300m.

$9000 = 2 \times 300^2 / L, \quad L = 20$ m

The frequency of this wavelength is 15MHz.

7.4 ANTENNA GAIN

To understand the concept of antenna gain, one needs to consider an "isotropic" antenna. This is a ficticious device which radiates power equally well in all directions. We say "ficticious" because one can never be built in practice. All antennas have a preferred direction of radiation. Still, we speak of isotropic antennas and their properties, because they make it easier to understand the properties of real antennas.

An important concept is that of radiated power density, expressed in watts/m^2 (or watts/cm^2). The radiated power from an antenna is spread across a broad expanse of space, and the power density is, roughly speaking, the total power divided by the total area.

For an isotropic antenna, the radiated power density simply depends on the transmitted power and the distance from the antenna.

$$P_d = P_t / 4\pi S^2 \qquad (7.2)$$

In other words, the power density is equal to the transmitted power divided by the total surface area of a sphere with a radius of S centered at the antenna.

The gain of an antenna is the ratio of the actual power density in the center of the antenna beam to the power density that would exist for an isotropic antenna. All real antennas have a gain greater than 1.0, or 0dB. Gain is usually expressed in dB, although its numeric value should be used in equations and calculations involving the value of gain.

The actual power density in the center of the beam for an antenna of gain, G, transmitted power P_t, measured at a distance S from the antenna, is

Power density $P_d = GP_t/(4\pi S^2)$ (7.3)

When an antenna is used as a receiver, the amount of power that it receives depends on two quantities. The first is the size and construction of the antenna itself, and the second is the power density of the incident electromagnetic wave. Of course, it is essential that the antenna be properly oriented to receive the microwave power, pointed at the source of the radiation, matched to the polarization of the wave, etc.

The physical area of the receiving antenna is called its "aperture" denoted by the symbol A. The power received by an antenna is

P_r = (Power density) x (Aperture) (7.4)

A good analogy is that of collecting rainwater. The intensity of rainfall could be described as R gallons per hour in a one square meter area. If you spread collecting pans with a total collecting area of A in square meters, then you would collect AxR gallons per hour of rainwater. In the microwave world, we have watts per square meter instead of rainfall, and antenna aperture instead of collecting area of buckets and pans.

Exercises

Use the above equations to calculate power density for the following problems.

(7.4) Calculate the power density (in watts/m^2) for an isotropic antenna at the following distances and transmitted power levels.

a. Transmitted power = 1W, distance = 5km
b. Transmitted power = 10W, distance = 10km
c. Transmitted power = 1kW, distance = 50km

Sample calculation

a. $P_d = 1/4\pi \times 5000^2 = 3.18 \times 10^{-9} W/m^2$

(7.5) What is the power density in the beam for an antenna with the following characteristics?

	Transmitted Power	Distance	Antenna Gain
a.	20 W	1.5 km	35 dB
b.	30 kW	500 km	12 dB
c.	1 W	50 km	3.5 dB
d.	10 mW	30 mi	55 dB
e.	15 W	20,000 mi	25 dB

Sample calculation

a. Gain = 35dB = 3000
$P_d = 3000 \times 20/4\pi \times 1500^2 = 2.1 \times 10^{-3} W/m^2$

(7.6) What antenna gain is required to achieve a power density of 1 $\mu W/m^2$ at a distance of 1,250 km from a 1 kW transmitter?

Calculation

By solving for gain G in Equation (7.3),

$$10^{-6} = G \times 1000/4\pi \times (1.25 \times 10^6)^2$$

we have $G = 1.96 \times 10^4 = 2 \times 10^4 = 43dB$.

(7.7) For a transmitted power of 150 W and an antenna gain of 20 dB, a) at what distance would the radiated power density in the main beam drop to a value of 800 nW/m^2? b) How far for a power density of 200 nW/m^2?

Calculation

a. $G = 20dB = 100$, We solve for the distance S_1 that gives a power density of 800nW/m^2.

$$800 \times 10^{-9} = 100 \times 150/4\pi S_1^2$$
we have
$$S_1 = \sqrt{1.49 \times 10^9} = 3.86 \times 10^4 m = 38.6km$$

b. We solve for the distance S_2 that gives a power density of 200 nW/m^2

$$200 \times 10^{-9} = 100 \times 150 / 4\pi S_2^2$$
we have
$$S_2 = 77.2 \text{ km.}$$

An alternative method is to recognize that the power density decreases as the square of distance. If the power density decreases by a factor of 4, the distance increases by a factor of $\sqrt{4} = 2$, i.e., twice the distance.

7.5 PARABOLIC ANTENNAS

Parabolic antennas are extremely common in microwave systems. Intuitively, one would expect that the aperture of a parabolic dish is equal to its area. In practice, dish antennas are only about 60% efficient due to loss of signal gathering power near the rim of the antenna. So, for parabolic antennas, a good rule of thumb is

$$A = 0.6\pi R^2 \tag{7.5}$$

Several useful formulas for gain and half power beamwidth (HPBW) are

$$G(dB) = 13.2 + 20 \log(f) + 20 \log(D) \tag{7.6}$$

$$HPBW(deg) = 57 (L/D) = 17.1/(fD) \tag{7.7}$$
and
$$G(dB) = 37.87 - 20 \log (HPBW) \tag{7.8}$$

Note that in Equations (7.7) through (7.8), f is in GHz and L an D are in meters.

(7.8) Calculate the aperture in square meters for the following parabolic antennas using Equation (7.5).

a. small 1 m diameter parabolic dish,
b. medium sized 3 m dish,
c. large, INTELSAT sized, 30 m dish.

Sample calculation

a. Radius of antenna = 0.5m, Aperture $A = 0.6\pi \times 0.5^2 = 0.47m^2$

(7.9) Calculate the ratio in dB between the total signal received by the INTELSAT dish in part (c) to the small antenna in part (a) of Exercise (7.8). Assume they are exposed to the same radiated power.

Calculation

We shall aproach the problem more mathematically. Consider two antennas of radii equal to R_1 and R_2 respectively. The corresponding apertures A_1 and A_2 are given by $A_1 = 0.6\pi R_1^2$ and $A_2 = 0.6\pi R_2^2$. We have,

$$A_1/A_2 = (R_1/R_2)^2$$

From the previous exercise, we set $R_1 = 15m$ and $R_2 = 0.5m$. For the same power density, the ratio of collected signal power is the same as the ratio of the antenna's apertures.

$$P_1/P_2 = A_1/A_2 = (R_1/R_2)^2 = (15/0.5)^2 = 900 = 29.5dB$$

(7.10) Communication satellites are 35,800 km above the equator. They typically transmit 5 W of power at 4 GHz from antennas with approximately 28 dB gain. Calculate the received power in dBm for each of the three examples in Exercise (8).

Sample calculation

a. The power density from the satellite to the surface of the Earth, given that the gain $G = 28dB = 631$, is

$$P_D = 631 \times 5/4\pi(3.58 \times 10^7) \ W/m^2 = 1.96 \times 10^{-13} \ W/m^2$$

Given that the aperture for a) is $0.47m^2$, the received power is

$$P = P_D \times A = 1.96 \times 10^{-13} \times 0.47 \ W = 9.2 \times 10^{-14} \ W = 92fW = -100.4dBm.$$

(7.11) (Note: more difficult) If the satellite antenna was 36 dB gain instead of 28 dB, how much transmitter power would be needed to receive the same power from a 1 m diameter antenna as that which the 30 m antenna received in Exercise (7.10)?

Calculation

The received power is a product of the power density from the transmitting antenna and the aperture of the receiving antenna. The power density of a transmitting antenna is $P_D = G \, P_t / 4\pi S^2$. For two arrangements to have the same received power (at the same distance), we have

$$G_1 \, P_{t\,1} \, A_1 = G_2 \, P_{t\,2} \, A_2$$

Setting $G_1 = 600$, $P_{t\,1} = 5$, $A_1 = 0.6\pi x15^2$, $G_2 = 4000$, $A_2 = 0.6\pi x0.5^2$, we solve for $P_{t\,2}$

$$600 \times 5 \times 0.6\pi \times 15^2 = 4000 \times P_{t\,2} \times 0.6\pi \times 0.5^2$$

$$P_{t\,2} = 675 \text{ W.}$$

(7.12) Calculate the antenna gain in dB for a 2.5 m parabolic antenna,

    a. at 4 GHz,
    b. at 12 GHz,
    c. at 30 GHz.

Sample calculation

    a. Using Equation (7.6), we have

      $G = 13.2 + 20 \log(4) + 20 \log(2.5) = 33.2 \text{dB.}$

(7.13) Calculate the HPBW (in degrees) for the same antennas as in Exercise (7.12).

Sample calculation

    a. Using Equation (7.7) for HPBW, we have

      $\text{HPBW} = 17.1/(4x2.5) = 1.71^\circ$

7.6 PATH LOSS AND THE LINK EQUATION

The most commonly used specification of an antenna's properties is the gain in dB. We calculated the received power from an antenna in terms of its aperture and the transmitter's gain. There is an equation which relates an antenna aperture to its gain and wavelength. It is

    $G = 4\pi\, A/L^2$ (7.9)

What this equation says is that an antenna's gain equals 4 times the aperture in "square wavelengths." Equation (7.9) can be solved for A to express the aperture in terms of the gain.

    $A = G\, L^2/(4\pi)$ (7.10)

Exercises

(7.14) Calculate the gain (in dB) for the following examples.

    a. Aperture = 50 m^2, L = 0.03 m
    b. Aperture = 250 m^2, L = 0.5 m
    c. Aperture = 60 m^2, L = 20 cm
    d. Parabolic dish, dia = 3.5 m, f = 4 GHz
    e. Parabolic dish, dia = 10 m, f = 4 GHz
    f. Parabolic dish, dia = 10 m, f = 14 GHz

(7.15) Calculate the aperture in square meters, given the gain

    a. Gain = 15 dB, L = 3 cm
    b. Gain = 28 dB, L = 15 cm

c. Gain = 45 dB, f = 44 GHz
d. Gain = 15 dB, f = 800 MHz

We can now combine everything we know about received power, antenna aperture, radiated power density and the relationship between aperture and gain to get the following equation. If we express all quantities in dB and dBm, we get

$$P_r = P_t + G_t + G_r - 120 - 20 \log(S) + 20 \log(L) \quad dBm \tag{7.11}$$

This can also be expressed in terms of the operating frequency as

$$P_r = P_t + G_t + G_r - 90.5 - 20 \log(S) - 20 \log(f) \quad dBm \tag{7.12}$$

where the operating frequency f is in GHz and the distance S in km.

(7.16) Calculate the received power in dBm at a site 30 km from a 20 W transmitter. the transmitter antenna gain is 35 dB, the receiving antenna gain is 25 dB and the operating frequency is 6 GHz.

Calculation

Using Equation (7.12), we have

P_t = 20W = 43dBm
P_t = 43 + 35+ 25 - 90.5 - 20 log(30) - 20 log (6) = -33.5dBm

(7.17) In Exercise (7.16), suppose we had to increase the received power by 8 dB. What are our options assuming we must maintain the same operating frequency ?

Calculation

From Equation (7.12), the parameters which have contribution to the received power are P_t, G_t, G_r, S, and f. Since the frequency f is kept fixed, the quantities that can be increased practically are the transmitted power P_t, the transmitter's gain G_t, and the receiver's gain G_r. The distance S can be decreased but not practically.

(7.18) Two identical antennas are used in a transmission link. Their gain is 35 dB, the operating frequency is 4 GHz, and they are 1 km apart. The transmitter operates at 1 W. What is the received power level? What is the received power if the antennas are only 200 m apart?

Calculation

$G_t = G_r$ = 35dB

P_r = 30 + 35 + 35 - 90.5 - 20 log(1) - 20 log(4) = -2.5dBm

8

RADAR BASIC

8.1 DEFINITIONS

The following symbols are used in this chapter.

f_d = Doppler frequency shift
f_o = transmitter frequency
v = velocity of the target relative to the radar
t = elapsed time
L = wavelength
R_{max} = the maximum range in meters
P_t = the transmitted power in watts
G = the transmitted antenna gain (numerical)
A_c = the radar cross section of the target in m^2
A_r = the aperture of the receiving antenna in m^2
P_{min} = the minimum detectable signal of the receiver in watts

8.2 DOPPLER RADAR

The simplest type of radar is a CW Doppler radar, which measures the velocity of a target by determining the frequency shift on the reflected signal. The amount of frequency shift is easily calculated. It is

$$f_d(Hz) = 3.43 \ v(mph) \ f_o(GHz) \tag{8.1}$$

or

$$f_d(Hz) = 3.90 \ v(Kts) \ f_o(GHz) \tag{8.2}$$

This frequency shift can also be written in terms of the wavelength of the radar signal

$$f_d(Hz) = 117 \ v(Kts) \ /L(cm) \tag{8.3}$$

Exercises

(8.1) A target is moving toward a radar at a velocity of 100 Kts. The radar is operating at 10 GHz. Calculate the Doppler shift.

Calculation

Using Equation (8.2), $f_d(Hz) = 3.90 \times 100 \times 10 = 3900$ Hz

(8.2) A 5 GHz radar receives an echo that is shifted up in frequency by 10 kHz. Calculate the target's velocity in mph.

Calculation

Using Equation (8.1) to solve for v(mph), we have

$$v(mph) = f_d(Hz) / [3.43 \times f_o(GHz)] = 10000 / [3.43 \times 5] = 583.1 \text{ mph}$$

(8.3) What Doppler shift will be measured by a 3 GHz radar tracking a target moving at 1,000 Kts toward the radar?

(8.4) If a 30 GHz radar measures a 200 kHz Dopler shift, what is the target's velocity in Kts?

(8.5) If a 50 kHz Doppler frequency is measured and a target's velocity is known to be 1282 Kts, calculate the radar transmission frequency.

Calculation

Using Equation (8.2) to solve for f_o, we have

$$f_o(GHz) = f_d(Hz) / 3.90 \times v(Kts) = 50000 / 3.90 \times 1282 = 10.0 \text{ GHz}$$

8.3 RADAR RANGE MEASUREMENTS

Many radars in use today use a pulse delay technique to determine the distance to the target. Since the velocity of the radar signal is well known, measuring the round trip time delay of the pulse easily yields the distance to the target. The radar signal travels at the speed of light, which is 3×10^8 m/s.

Since a pulse makes a round trip to the target and back, the range to the target is only half the distance traveled by the pulse, or

$$\text{Range} = (1/2) \text{ (speed)} \times \text{(time)} \qquad (8.4A)$$

Using kilometers and microseconds,

$$R(km) = 0.15t \text{ } (\mu s) \qquad (8.4B)$$

In nautical miles, Equation (8.4) becomes

$$R(Kts) = 0.081t \text{ } (\mu s) \qquad (8.5C)$$

(8.6) What is the signal delay time for a search radar and a target at 450 nautical miles?

Calculation

Using Equation (8.5C) to solve for t (μs), we have

$$t \text{ } (\mu s) = R \text{ (Kts)} / 0.081 = 450 / 0.081 = 5555 \text{ } \mu s = 5.56 \text{ms}$$

(8.7) (Note: more difficult) Long rang search radars usually rotate very slowly. Why?

8.4 THE RADAR EQUATION

The basic radar equation determines how far away a target can be detected for a given radar system given the following parameters: antenna gain, transmitted power, target cross section, receiver antenna aperture, and receiver sensitivity.

The basic equation reads

$$R_{max} = \{\frac{P_t\,GA_r\,A_c}{(4\pi)^2\,P_{min}}\}^{1/4}$$

(8.6)

where the quantities are defined in Section 8.1.

The equation is rather involved, but one important factor stands out. The range is directly proportional to the transmitted power, the transmitter gain, the receiver aperture, and the target cross section, and inversely proportional to the minimum detectable signal, as one would expect. Mathematically, the range only varies as 1/4 power of these quantities. That means, it takes a sizeable change in any of these parameters to make a significant change in R_{max}.

For example, to double the range of a radar, we must increase the transmitter power by 16 times (2^4). See the following exercises for more illustrations of this principle.

Equation (8.6) can be solved for one of the unknown parameters, given that the others are known.

(8.8) Find the maximum range of a radar from the following parameters. Power = 100 kW, Gain = 30 dB, Receiver aperture = 10 m^2, Minimum detectable signal = 0.1 nW, Target cross section = 1 m^2

Calculation

Using Equation (8.6) to find the maximum range, we have

$$R_{max} = \sqrt[4]{\frac{10^5 \times 10^3 \times 10 \times 1}{(4\pi)^2 \times 10^{-10}}}\,\text{m} = \sqrt[4]{6.33 \times 10^{16}}\,\text{m}$$

$$= 1.58 \times 10^4\,\text{m} = 15.8\text{km}$$

(8.9) If the transmitter power in Exercise (8.8) is increased to 500 kW, what is the new range?

(8.10) Suppose the target in Exercise (8.8) has a cross secton of 25 square meters, what is the maximum detectable range?

(8.11) For the same radar in Exercise (8.8), if we increase the antenna gain to 40 dB, what is the absolute minimum transmitter power we can use to detect a target with a cross section of 0.2 square meters at a range of 20 km?

(8.12) If a 5 Watt transmitter is available, how large a target can be detected at a range of 10 km? Use the same parameters as the previous exercises.

9

COMMUNICATION SATELLITE LINKS

9.1 DEFINITIONS

S = 22,300 miles, the approximate transmission line distance to receivers in the United States and Europe.
W = the diameter in miles of the satellite footprint.
HPBW = half power beam width in degrees
EIRP = satellite EIRP in dBW
G/T = receiver G/T in dB/K
A = path atteuation in dB
k = Boltzmann's constant in decibel form = -228.6 dBW/(Hz K)
B = Satellite bandwidth in dB-Hz. Calculate this by taking 10 log(bandwidth in Hz)

9.2 INTRODUCTION

This chapter is divided into several sections. Each section deals with a specific aspect of communication satellites. The last section presents several exercises to calculate the overall performance of a typical satellite link.

9.2 SATELLITE ANTENNAS

For many applications the down-link is the limiting factor in the system performance, because the transmitting power from the satellite is only 5 to 10 watts per transponder due to the size of the solar power generators. However, the ground station transmitters can be made quite powerful since power sources are plentiful.

The primary requirement for the satellite antenna is to focus the beam on the portion of the earth's surface where the receiving antennas are located. The satellite does not broadcast at a particular antenna (except perhaps in specialized military systems), but it beams its signals to a broad geographic area where many receiving antennas are located.

Useful formulas describing the important properties of parabolic antennas can be found in Chapter 7, Equations (7.6) - (7.8). For purposes of calculation, the HPBW of a space antenna which will cover an area (footprint) of diameter W will be

$$HPBW = 57 \ (W \ /S) \ degrees \tag{9.1}$$

where W and S are given in Section 9.1.

Exercises

(9.1) Calculate the antenna beamwidth and gain (in dB) for a satellite antenna which must cover

a. the entire United States (3000 miles)
b. New England (300 miles)
c. a major metropolitan area (50 miles)

Sample calculation

a. Using Equation (9.4), we have HPBW = 57x3000 /22300 = 7.67^{o}

Using Equation (7.8), G = 37.87 - 20 log(7.67) = 20.2 dB.

(9.2) An advanced satellite concept is to use satellites for mobile communications with "spot" beams which have very narrow beamwidths. In one such proposal, satellites would generate beams with a footprint that is 250 miles in diameter, operating at approximately 800 MHz. Calculate the HPBW, the gain, and the diameter of the satellite antenna required for such a system.

Calculation

From Equation (9.1), we have HPBW = 57x250/22300 = 0.64^{o}

From Equation (7.8), we have

G(dB) = 37.87 - 20 log(0.64) = 37.87 - (-3.9) $\approx$ 41.8dB

Using Equation (7.7) to solve the antenna diameter D, we have

D = 17.1 / f x HPBW = 17.1 / 0.8 x 0.64 = 33.4m

(9.3) Secure communications is an imortant consideration for military systems. A drawback of satellite systems is that the beamwidths of the spacecraft antennas are too large to focus on individual receivers. A typical antenna projects a beam with a footprint hundreds of miles wide.

a. What two parameters should be changed to reduce the size of an antenna's footprint?
b. To obtain a footprint that is only 10 km in diameter, what size satellite antenna would be necessary operating at 4 GHz? What size at 20 GHz? Does either of these appear to be feasible?

Calculation

a. Increase either the operating frequency or the antenna size.

b. Given: Footprint W = 10km = 5.68 miles, we can find the HPBW from Equation (9.1)

HPBW = 57x5.68/22300 = 0.0145^{o}

Using Equation (7.7) to solve for the antenna diameter D, we have.

D(m) = 17.1/fxHPBW

For f = 4 GHz, D = 17.1/4x0.0145 = 294.36m (impractical)
For f = 20 GHz, D = 17.1/20x0.0145 = 58.87m (reasonably practical)

9.4 EARTH TERMINAL ANTENNAS

As discussed in Section 15.3 of the reference text, the primary figure of merit for a transmitting system is the effective isotropic radiated power (EIRP) which is the decibel sum of the transmitted power in dBW and the antenna gain in dB.

$$EIRP \text{ (dBW)} = P_t \text{ (dBW)} + G_t \text{ (dB)} \tag{9.2}$$

Units of dBW are explained in Chapter 1. They are the ratio in dB of the actual power to 1 Watt.

For a receiving antenna, the important quantity is the ratio of the receiver gain to the equivalent noise temperature T_e, the ratio is known as G/T.

The equivalent system noise temperature is

$$T_e = T_a + T_r \tag{9.3}$$

i.e., the sum of the equivalent antenna temperature and the equivalent temperature of the receiver electronics.

Usually $T_a \approx 60$ K. If the earth terminal is quite far north so that the antenna must be aimed quite close to the horizon, T_a will be 5 to 10 K higher. T_r ranges between approximately 65 K and 120 K, depending on the price of the unit and the requirements of the receiving system.

(9.4) Calculate the G/T of a 10 m antenna with a low noise amplifier (LNA) temperature of 85 K and an antenna temperature of 65 K.

Step 1. The equivalent system noise temperature $T_e = T_a + T_r = 65 + 85 = 150$ K
Step 2. Use Equation (7.6) to calculate the antenna gain

$$G = 18.3 + 20 \log(4) + 20 \log(10) = 1.08 \times 10^5.$$

Step 3. Divide the antenna gain by the system noise temperature to get

$$G/T = 720 \text{ K}^{-1} = 10 \log(720) = 28.6 \text{ K}^{-1}.$$

(9.5) Calculate G/T in dB for the following antennas and LNAs. Assume an antenna noise temperature of 65 K.

Antenna Diameter (m)	LNA Noise Temperature (K)		
	60	90	120
1.5			
6.0			
10.0			
30.0			

9.5 LINK CALCULATIONS

The parameter that determines the satellite link performance is the carrier to noise ratio at the receiver, C/N. In practical systems this is determined by two quantities, the EIRP of the satellite, and the G/T of the receiver.

The carrier to noise is expressed in dB and is

$$C/N \text{(dB)} = EIRP + G/T - A - k - B \tag{9.4}$$

where the quantities are given in Section 9.1.

Since the transmission distance for a geosynchronous satellite is essentially fixed, the path attenuation only depends on the transmit frequency and is given by

$$A(dB) = 183.9 + 20 \log(f) \qquad\qquad (9.5)$$

where f is in GHz.

In the exercises that follow, all the information is provided. You have to determine how to calculate the two parameters, EIRP and G/T to determine the down link C/N.

(9.6) A spacecraft transmits 5 W (7dBW) of power at 4 GHz with a 24 dB gain antenna. The transmission bandwidth is 36 MHz. On the ground, a 10 m diameter antenna is operating with an LNA temperature of 85 K, and an antenna temperature of 65 K. Calculate the C/N(dB) at the input to the receiver.

Step 1. Use Equation (9.2) to calculate the EIRP. EIRP = 24 + 7 = 31 dBW.

Step 2. Calculate the G/T. This is worked out in Exercise (7.4) for the same antenna,
$G/T = 28.6\ K^{-1}$.
Step 3. Calculate the bandwidth. $B = 10 \log(36 \times 10^6) = 75.6$ dB Hz.
Step 4. Put it all together

$$\begin{aligned} C/N &= EIRP + G/T - A - k - B \\ &= 31 + 28.6 - 195.9 - (-228.6) - 75.6 = 16.7\ dB \end{aligned}$$

(9.7) A domestic commercial satellite broadcasts at 4 GHz to the eastern half of the US with a transmitted power of 5 W and a bandwidth of 40 MHz. The footprint has a diameter of 1,500 miles. What is the receiver C/N in dB if the G/T of the receiver is $19.4\ K^{-1}$.

(9.8) For the same satellite as in Exercise (7.7), an earth terminal uses a 3 meter antenna and a 120 K LNA. The antenna noise temperature is 70 K. What is the received C/N?

(9.9) If the 120 K LNA in Exercise (7.8) was replaced with a 70 K version, how much improvement in C/N could be expected?

9.6 DBS AND TVRO SYSTEMS

TVRO stands for television receive only and describes the most widespread type of satellite receiver in operation today. By 1985, approximately 1.5 million C-band (4 GHz) earth stations had been installed worldwide to receive television broadcasting from domestic satellites. The parabolic dishes are typically 2.5 to 3.5 m in diameter and are installed in backyards, parking lots, or wherever room can be found.

A competing concept for consumer TV reception from satellites is called DBS (direct broadcast satellite). This system is planned to operate at 12 GHz and will use much smaller parabolic antennas, typically mounted on rooftops. The advantage of operating at 12 GHz is that the receiving antennas are smaller for the same gain. However, the receiver noise temperature, T_r, is higher because LNAs have poorer performance quality at 12 GHz than at 4 GHz.

(9.10) The purpose of this exercise is to compare the performance of DBS and TVRO systems. A typical TVRO and DBS links have the following parameters. Assume that the receiver signal to noise requirements for both systems are identical. Use the link Equation (9.5) and remember to use Equation (9.5) to calculate a new path attenuation for DBS due to the higher frequency of operation.

TVRO:

Satellite transponder P_t = 7 W, Satellite antenna gain = 28.5 dB,

Earth station antenna diameter = 3 m, T_r = 80 K, T_a = 65 K, f = 4GHz, and B = 4MHz.

DBS:

Satellite transponder P_t = 200 W, Satellite antenna gain = 28.5 dB,

Earth station antenna diameter = 1 m, T_r = 400 K, T_a = 65 K, f = 12 GHz, and B = 4MHz.

a. What is the received C/N in dB for the TVRO link?

b. What DBS transponder output power is required to achieve the same C/N?
c. How can the DBS system work with such a small ground antenna?
d. Assume that the receiver signal to noise requirement for both systems are identical. Use the link equation, Equation (9.4), to calculate a new path loss using for DBS due to the higher frequency of operation.

9.7 VSAT Ku BAND TERMINALS

Another recent evolution in satellite systems is the emergence of low cost VSAT (very small aperture terminals) Ku-band earth terminals for use in low speed data communications systems. The appeal of these systems is that many business users of data communication links do not need high speed. Applications such as linking a remote terminal with a computer, transmitting orders, sending stock prices, updating inventory lists, etc, can use data rates of only a few kilobits per second.

The low cost arises because the dishes are very small, less than three feet in diameter. The overall cost of the earth station is less than $10,000. This compares quite favorably with high data rate systems with earth terminals which cost several hundred thousand dollars and antenna diameters of 10 m or more.

(9.11) (Note: more difficult) To see how the small antenna sizes are possible, calculate the carrier to noise ratio at the receiver for a hypothetical VSAT system with the following parameters.

Operating frequency = 12 GHz
EIRP = 31.0 dB
Ground antenna diameter = 0.7 m
Antenna temperature = 65 K
Receiver temperature = 400 K
Bandwidth = 20 kHz

Compare the result with the Exercise (9.6). Notice that, although the G/T of the VSAT receiver is much lower than before, the received C/N is nearly the same. Why is this so?

SOLUTIONS TO PROBLEMS

CHAPTER ONE

1.1

frequency f(Ghz)	free space wavelength Lo(cm)
10.0	3.0
7.2	4.167
44.7	0.667
2.5	**12.0**
0.5	**60.0**
2.0	**15.0**

1.2

Number	dB
12	16
15.4	35
19.5	88.5
0.02	-17
$5.0x10^{-7}$	-63
$3.0x10^{-5}$	-45
16	**12**
500	**27**
$7x10^6$	**68.5**
$1.6x10^{-5}$	-48
$4.0x10^{-10}$	-94
0.007	-21.5

1.3

power	dBW	dBm
5mW	**-23**	**7**
35W	**15.4**	**45.4**
6kW	**38**	**68**
20μW	**-47**	**-17**
0.5pW	**-123**	**-93**
7nW	**-81.5**	**-51.5**
3MW	**65**	**95**

1.4

dBm	mW	W
19	**80**	**0.08**
37	**5000**	**5**
4	**2.5**	**$2.5x10^{-3}$**
52	**$1.6x10^5$**	**160**
-26	**$2.5x10^{-3}$**	**$2.5x10^{-6}$**
-78	**$1.6x10^{-8}$**	**$1.6x10^{-11}$**

1.6   -22 dBm

1.7   **23 dB**

1.8   **52 dB**

1.10   -17 dBm

1.11   28 dB

1.13   10 dB

1.14   24 dBm

1.18   42 dBm

1.20   -19 dBm

1.21   -13 dBm

1.25 Isolator forward loss

forward loss	isolation	device RL	effective RL
0.2	25	30	**55.2**
0.6	35	10	**45.6**
1.5	30	10	**41.5**

(1.26)

RL(dB)	RL(#)	Δ	VSWR
5	**3**	**0.58**	**3.74**
25	300	**0.058**	**1.12**
14	25	**0.2**	1.5
1.94	**1.56**	0.8	**9**

CHAPTER TWO

(2.2)

f(GHz)	Skin depth in μm
1.5	$2/\sqrt{1.5} = 1.63$
2.5	$2/\sqrt{2.5} = 1.26$
5.7	$2/\sqrt{5.7} = 0.84$
10.3	$2/\sqrt{10.3} = 0.62$
15.2	$2/\sqrt{15.2} = 0.51$
25.0	$2/\sqrt{25} = 0.4$

(2.3)

f(GHz)	L_o(cm)	L_g(cm)
1.0	$30/1 = 30$	$30/\sqrt{2.2} = 20.23$
2.5	$30/2.5 = 12$	$12/\sqrt{2.2} = 8.09$
4.8	$30/4.8 = 6.25$	$6.25/\sqrt{2.2} = 4.21$
6.0	$30/6 = 5$	$5/\sqrt{2.2} = 3.37$

(2.4)

f(GHz)	L_O(cm)	L_g(cm)
1.0	30	$30/\sqrt{5.3} = 13.03$
3.5	8.57	$8.57/\sqrt{5.3} = 3.72$
5.7	5.26	$5.26/\sqrt{5.3} = 2.27$
6.2	4.84	$4.84/\sqrt{5.3} = 2.10$

(2.5)

f(GHz)L_O (cm)		L_g(cm)
10.0	3	1.3
9.0	0.011	0.0048
8.5	3.53	1.53
8.0	3.75	1.63
5.0	6.0	2.61

2.10 $C = 16.8$ pF/ft
        $L = 128.6$ nH/ft

2.11

Point	Normalized value	Nature	Actual value (ohm)
P	0.15+j0.55	R&L	11.25+j41.25
Q	0.55-j0.50	R&C	41.25-j37.5
R	2.0+j0.5	R&L	150+j37.5
S	0.15-j0.95	R&C	11.25-j71.25
W	0.5+j2.0	R&L	37.5+j150
X	0.3-j2.1	R&C	22.5-j157.5

2.12 2.27×10^8 m/s

CHAPTER THREE

3.4 P = 0.15 + j 0.55; Q = 0.55 - j 0.5; R = 2.0 + j 0.5; S = 0.15 - j 0.95; W = 0.5 + j 2.0;
     X = 0.3 - j 2.1

3.9   VSWR = 6.7, $\Delta = 0.74$ RL = 2.5 dB

CHAPTER FOUR

4.2   20mA, 50mA, 10mA, 5mA, 60mA (all from A to B)

4.3   7mA(A to B), 20mA(A to B), 5mA(B to A), 30mA(B to A), 20mA(B to A)

4.5   1mA, 10mA, 30mA, 10mA, 20mA

4.6   7mA, 40mA, -4mA, -20mA, 23mA, -15mA

4.7

I_1(mA)	I_2(mA)	I_3(mA)	I_4(mA)
10	5	**8**	7
20	2	10	**12**
8	**5**	6	7
6	50	32	24
10	6	0	**16**
8	0	**5**	3
0	16	7	**9**
12	8	**5**	15
42	3	15	30
36	5	21	20

4.9

V_{CC}	R_C	V_{CE}	I_C	I_1	I_2	I_B	V_{BE}	R_1	R_2
20	**375**	5	40	**2.4**	$5I_B$	**0.4**	1	**7.9k**	500
18	**400**	6	30	**2.7**	$8I_B$	**0.3**	1	**6.3k**	416
25	500	**5**	40	**3.2**	$7I_B$	**0.4**	1.5	**7.3k**	535
15	**350**	8	**20**	$10I_B$	**1.8**	0.2	1.2	**6.9k**	666
20	**1k**	10	10	$8I_B$	**0.7**	**0.1**	0.7	**24.1k**	**1k**

4.10

V_{CC}	R_C	V_{CE}	V_E	R_E	I_C	I_B	I_1	I_2	V_{BE}	R_1	R_2
25	**500**	10	5	**248**	20	0.2	**2**	$9I_B$	1	**9.5k**	**3.3k**
20	200	**11**	3	**99**	30	0.3	$10I_B$	2.7	1.5	**5.2k**	**1.7k**
18	**124**	7	6	150	**40**	0.4	$8I_B$	2.8	1	**3.4k**	**2.5k**
20	**400**	8	4	**198**	20	0.2	**1.2**	$5I_B$	1.5	**12k**	**5.5k**

4.11

S_{11}	S_{21}	S_{22}	S_{12}	RL(dB)	Input VSWR	Gain (loss)dB	Forward RL(dB)	Output VSWR	Reverse Gain(isol.)dB
0.4	2.0	0.25	0.2	8	**2.33**	**26 Gain**	12	1.67	**14 isol**
0.1	0.1	0.1	0.1	20	**1.22**	**20 loss**	20	1.22	**20 loss**
0.2	0.89	0.15	0.05	14	**1.5**	**1 loss**	16.5	1.35	**26 isol**

a: 26dB amplifier, b: 20dB attenuation, c: isolator with 1dB loss and 26dB isolation.

4.16

0.995nH, 1.99nH; 3.32nH, 1.19nH; 0.199nH, 0.796pF; 0.398pF, 0.531nH

4.18

V_{GS}(V)	I_{DS}(mA)	V_1(V)	V_D(V)	V_{DS}(V)
-1	**35**	1	10	**10**
-2	15	2	8	**8**
-3	5	3	5	5
-2.5	**10**	2.5	6	6
-1.75	20	**1.75**	4	4

4.20

$V_{GS}(V)$	$I_{DS}(mA)$	$R_S(\Omega)$	$V_D(V)$	$V_{DS}(V)$
-1	35	**28.6**	5	**4**
-2	15	**133.3**	10	**8**
-1.5	50% of I_{DSS}	**60**	12	**10.5**
-3	10% of I_{DSS}	**600**	8	**5**
-1.75	20	**87.5**	10	8.25
-1.25	30	**41.7**	7	5.75
-2.5	10	**250**	10	7.5
-0.5	40	**12.5**	5	4.5

4.22

a.

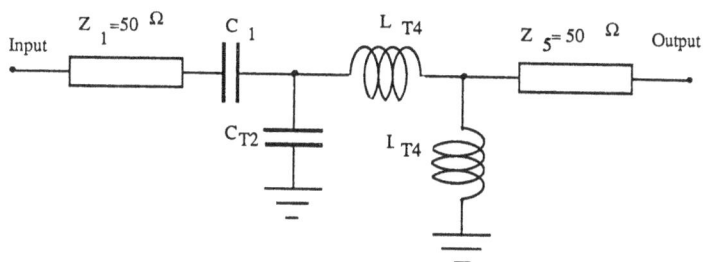

Prob. 4.22A

b.

Answer:

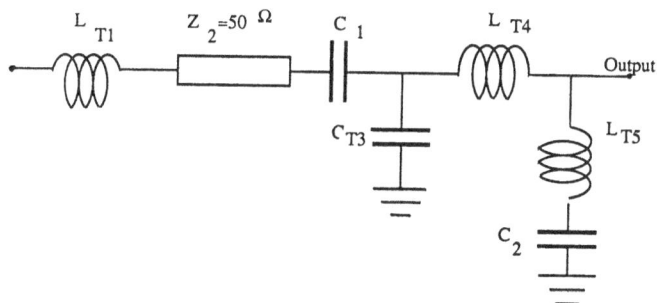

Prob. 4.22B

c.

Answer

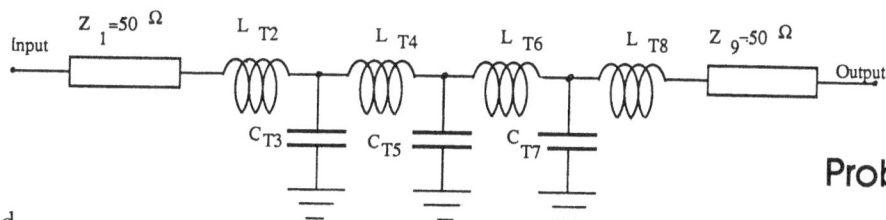

Prob. 4.22C

d.

Answer:

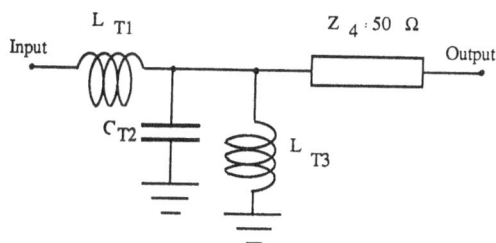

Prob. 4.22D

4.23   9.56, 5.78, 9.10, 7.53, 18.3 GHz

4.24   7.65, 4.04, 3.30, 6.60, 16.8 GHz

4.25   5.03, 2.52, 1.01, 1.59, 10.1 GHz

4.26   1pF, 7.12GHz; 2pF, 6.16GHz; 3.5pF, 5.71GHz; 0.5pF, 14.2GHz; 1.5pF, 11.6GHz; 2.5pF, 11.0GHz

CHAPTER FIVE

5.1   a.

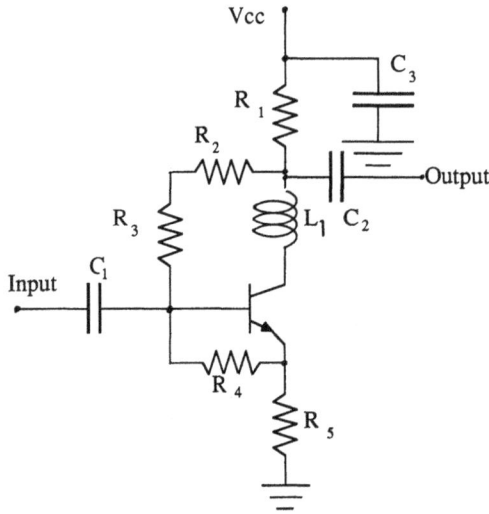

Prob. 5.1A

b. R_1, R_2, R_3, R_4, R_5
c. R_2, R_3
d, L_1
e. negative feedback would increase and gain would decrease
f. R_5 provides negative feedback through the emitter circuit. If R_5 were shorted the negative feedback would decrease thereby increasing the gain. Note that R_5 is generally a small value in the range of 5Ω to 20Ω.

5.2.   a.   Answers

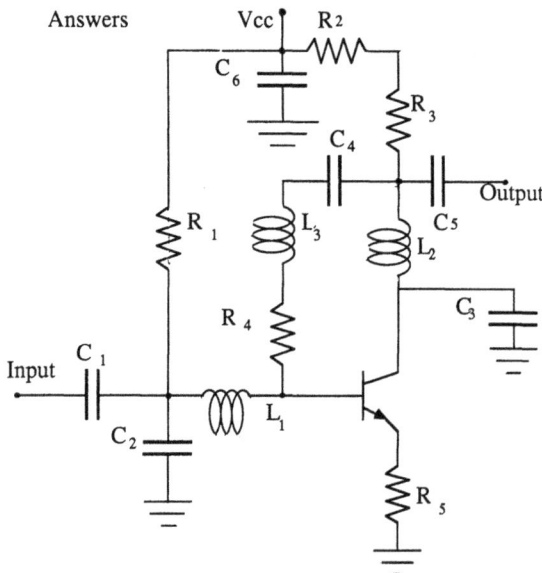

Prob. 5.2A

b. R_1, R_2, R_3, R_5

c. L_1, C_2

d. L_2, C_3

e. C_4, L_3, R_4

f. The reactance of a series inductor is directly proportionally to the frequency of operation. As the frequency of the amplifier decreases, the series reactance of L_3 will decrease which has the effect of increasing the negative feedback. The gain will decrease as the frequency decreases. The ultimate purpose of L_3 is to compensate the open loop gain of the transistor through the use of frequency dependent negative feedback to produce a flat gain amplifier over a wide frequency range.

5.3.

a.

Prob. 5.3A

b. YIG sphere, coupling loop, and static magnetic field

c. L_{T2}, C_{T3}

d. To provide a negative dynamic resistance which produces the unstable condition

e. 100%. Note that the gate and source are DC ground making $V_{GS} = 0V$. When

$$V_{GS} = 0V, I_{DS} = 100\% \text{ of } I_{DSS}.$$

5.4.

a.

Prob. 5.4A

b. YIG sphere, coupling loop, and static magnetic field

c. R_1, R_2, R_3, R_4

d. $C_{T1}, L_{T2}, L_{T3}, L_{T4}$

e. To provide a negative dynamic resistance which produces the unstable condition

f. T_1 would be lengthened thereby increasing the effective shunt capacitance of C_{T1}.

5.5.   a.

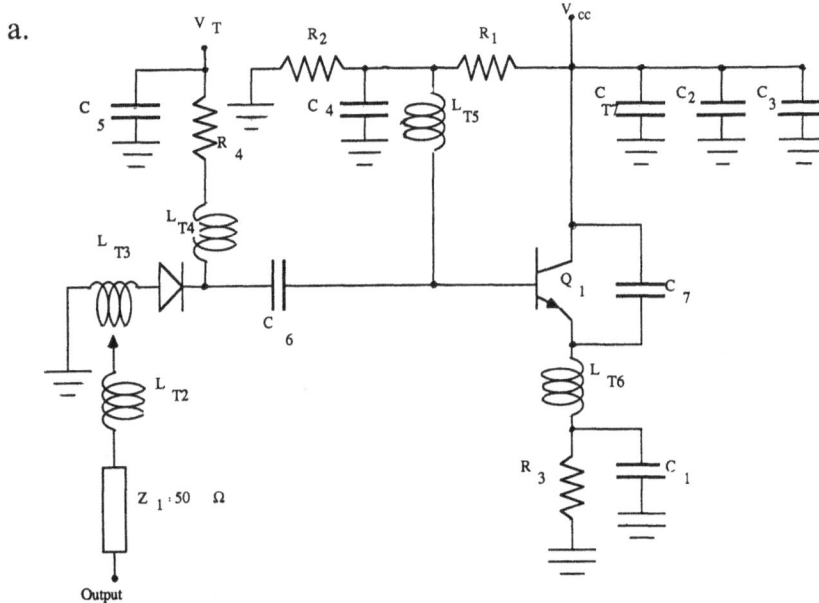

Prob. 5.5A

b. R_1, R_2, R_3

c. L_{T3}, C_V of the varactor diode, and C_t of the transistor. Circuit is a lumped element resonator.

d. To produce a negative dynamic resistance and therefore an unstable condition.

e. L_{T2} taps inductor L_{T3} which is part of the resonator

f. By moving the connection of L_{T2} along L_{T3} in a direction away from ground or toward the varactor diode.

5.6.   a.

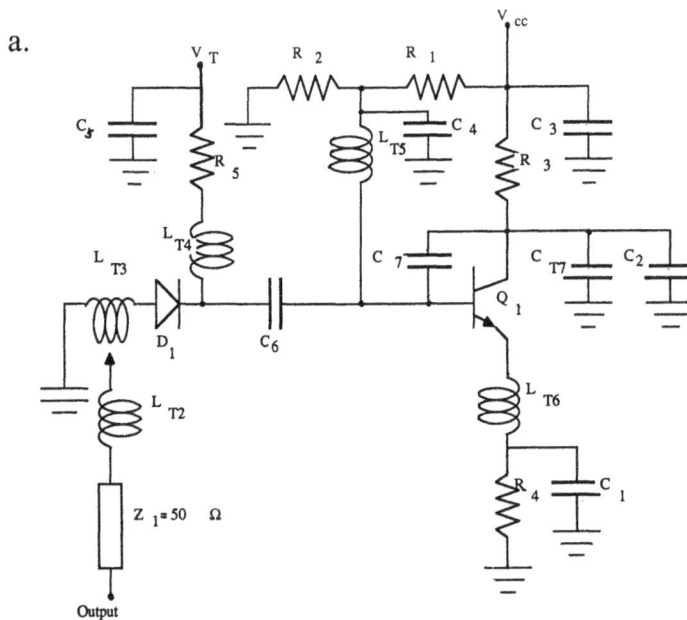

Prob. 5.6A

b. R_1, R_2, R_3

c. L_{T3}, C_V of the varactor diode, and C_t of the transistor

d. Reverse biased

e. Decrease

f. The output frequency will increase

5.7   a.

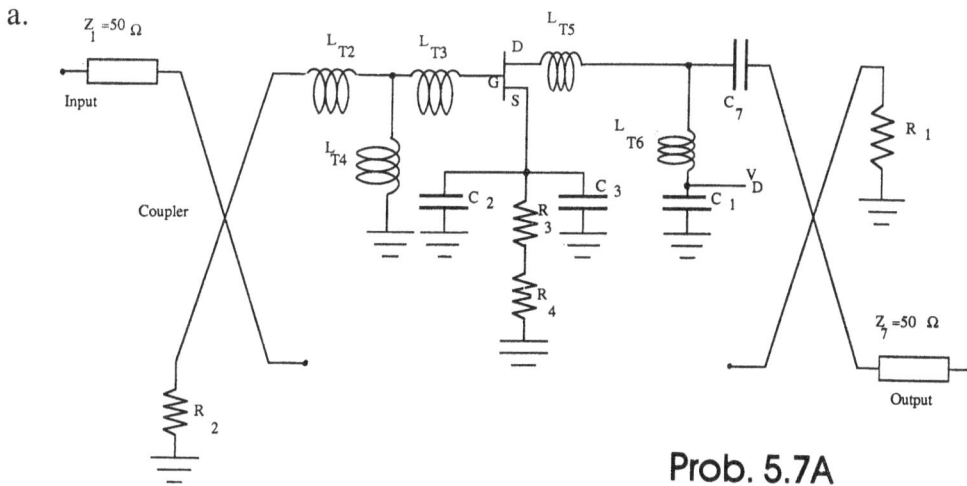

Prob. 5.7A

b. L_{T2}, L_{T3}, L_{T4}

c. L_{T5}, L_{T6}, C_7

d. self bias

e. To provide a good VSWR over wide bandwidths so that the circuits can be cascaded.

f. I_{DS} will increase because V_{GS} becomes less negative.

g. The effective inductance of L_{T3} will decrease.

5.8   a.

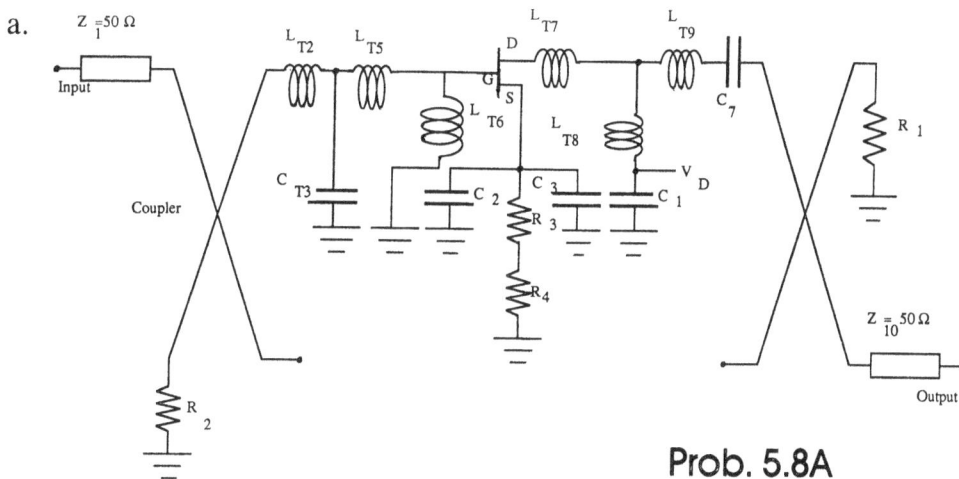

Prob. 5.8A

b. $L_{T2}, C_{T3}, L_{T5}, L_{T6}$

c. L_{T7}, L_{T8}, L_{T9}

d. The effective capacitance of C_{T3} would increase.

e. Decrease the source resistance by shorting R_3 or R_4

f. Since the signals at T_2 and T_{11} are 90° out of phase, their reflected signals will also be 90° out of phase. These signals combine at R_2 and are absorbed such that the input appears to have a good VSWR.

g. Since V_{GS} would be 0V, $I_{DS} = I_{DSS}$.

CHAPTER SIX

6.2   -157; -138; -127; -108; -98.4; -87; -74 dBm

6.5   7,975; 598; 72,500; 1.27×10^7; 109; 1,450 K

6.7 $NP(dBm) = P_{in} - S/N = -58dBm = 1.59nW$
The source temperature from Equation (6.3B) is $T_S = T_0 NP / NP_0 B = 2.87 \times 10^6$ K. Since T_S is not 290K, so Equation (6.5) cannot be used.

6.8   27.1dB; -6.6dB

6.9 $NP(dBm) = P_{in} - (S/N)_{in} = 3 - 46 = -43dBm = 5 \times 10^{-8} W$.
Use Equation (6.3B) to find $T_S = 9.06 \times 10^7 K$.
From Equation (6.6), solve for $(S/N)_{out} = 3.88 \times 10^4 = 45.9dB$.

6.10   2.37pW(-86.2dBm) ; 2.12pW(-86.7dBm); 75.4pW(-71.2dBm); 7.54nW(-51.2dBm)

6.11 4.95×10^{-3}; 61.7; 846 pW

6.12   75.1K; 4.4dB; 7.9dB; 14,244K; 289K; 0.5dB; 0dB(the amplifier adds no noise)

6.13   12.0; 2.42; 9.35dB

CHAPTER SEVEN

7.1   240; 8,000; 7,200 m

7.2   300; 1,200 km

7.4   7.96; 31.8 nW/m^2

7.5   151nW/m^2; 71.3pW/m^2; 108nW/m^2; 0.364pW/m^2

7.7   38.6; 77.3 km

7.8   4.24; 424 m^2

7.9   -90.8; -70.8 dBm

7.12   42.7; 50.7 dB

7.13   0.57; 0.23 degree

7.14 58.4; 41.0; 42.8; 43.3; 50.2; 61.1 dB

7.15   22.6cm^2; 1.13m^2; 0.117m^2; 0.354m^2

7.18   S= 200m; P_r = 11.4dBm

CHAPTER EIGHT

8.3   11.7kHz

8.4   1,790Kts

8.7   Long range radar rotates slowly to allow time for the pulse to return from a distant target.

8.9   23.6km

8.10   35.3km

8.11   Use Equation (8.6) to find P_t, $P_t = 1.26 \times 10^5$ W

8.12   Use Equation (8.6) to find A_c, $P_c = 315.8 \text{m}^2$

CHAPTER NINE

9.1 0.77^o, 40.2dB; 0.13^o, 55.8dB

9.5

Diamter	LNA temp		
	60	90	120
1.5	7.8	6.9	6.1
6.0	19.8	18.9	6.1
10.0	24.3	23.3	22.6
30.0	33.8	32.9	32.1

9.7   HPBW= 3.83^o, G= 26.2dB, EIRP= 33.2dBW, C/N = 9.3dB

9.8   The only difference is G/T. With 3m antenna, G= 34.8dB= 3009.
From Equation (7.6), G/T = 3009/(120+70) =15.8 , 10log (15.8) = 12 K^{-1} which is 4.4
less than Exercise (9.7). Therefore, C/N = 9.3 - 4.4 = 4.9dB

9.9   From Exercise (9.8), G/T = 3009/(70 + 70) $\doteq$ 21.5, 10log(21.5) = 13.3K^{-1}, a 1.3dB
improvement since all of the other parameters are the same.

9.10   a. EIRP = 37.0dBW, Rec. G = 13.2 + 20log(3) + 20log(4) = 34.8dB = 3009,
          G/T = 10log(3009/145) = 13.2 K^{-1}, C/N = 16.9dB
       b. Rec. G = 34.8dB = 3009, G/T = 10log(3009/465) = 8.1K^{-1}, P_t = 23.2dBW = 209W

       c. The DBS system works because the satellite transmitted power is much higher than that
          of TVRO. The receiver antenna can be smaller and yet have the same gain because the
          frequency is higher than TVRO.

9.11   Rec. Gain = 13.2 + 20log(0.7) + 20log(12) = 31.7dB = 1474
       G/T = 10 log(1474/465) = 5.0 K^{-1}, C/N = 31 + 5.0 - 205.5 + 228.6 - 10log(20×10^3) =
       16.1dB
       The C/N values are nearly equal because the bandwidth in the VSAT system is much lower.

9.12   G/T = 28.4K^{-1} = 691.8

NOTES

NOTES

NOTES

www.ingramcontent.com/pod-product-compliance
Lightning Source LLC
Chambersburg PA
CBHW081546220326
41598CB00036B/6581

COOPERATIVE EVOLUTION

RECLAIMING DARWIN'S VISION